Joshua D. Wachs, Busy Brain Studios

NICCO MELE is a leading forecaster of business, politics, and culture in our fast-moving digital age. Named by *Esquire* magazine as one of America's "Best and Brightest," he served as webmaster for Howard Dean's 2004 presidential campaign and popularized the use of technology and social media for political fund-raising, reshaping American politics. Not long after, he helped lead the online efforts for Barack Obama in his successful bid for the U.S. Senate. Mele's firm, EchoDitto, is a leading Internet strategy company working with nonprofit organizations and Fortune 500 companies, among them Google, AARP, the Clinton Global Initiative, the United Nations, and others. He also serves on a number of boards, including the Nieman Foundation for Journalism at Harvard, is a cofounder of the Massachusetts Poetry Festival, and is on the faculty at Harvard's Kennedy School of Government.

Visit his Web site, NiccoMele.com, and follow him on Twitter @nicco.

Additional Praise for

THE END OF BIG

"An important read for anyone curious about what the future might look like . . . the end of big is hitting many aspects of our lives. And Mele makes us seriously think about the world we live in today and, more important, how we'd like to live in it tomorrow."   —*Fortune*

"Anyone who is not asleep is unsettled by the speed and sweep of technological change, as it upends our workplace, our institutions, leisure, culture, individual and communal behavior. To comprehend the awesome changes we have and will experience, the opportunities and the pitfalls, Nicco Mele's *The End of Big* is a clear-eyed, compellingly written account, bursting with vivid anecdotes and analysis."

—Ken Auletta, *The New Yorker* writer and author of *Greed and Glory on Wall Street, World War 3.0,* and *Googled*

"In *The End of Big*, forecaster Nicco Mele—one of the Internet's early masters—looks technology squarely in the eye and asks the hard questions: Exactly how powerful is our new-found connectivity, and what's its effect on the media? On politics and government? On business? And on our culture? If you want to know what's really going on, get this book—and see the future and your options with new eyes."

—Alex Castellanos, political consultant and media commentator

"The intense and direct way the Internet and smart mobile devices connect us and the planet challenges existing institutional arrangements everywhere we look. *The End of Big* presents a provocative analysis of a world on the cusp of disruptive change and asks if we have the vision and will to remake it along small-d democratic lines."

—Mitch Kapor, founder of Lotus Development Corp.

"Is it news that we all now live in a constantly connected brave new world? No. But what is news is that technology has become an accelerating force in its own right. Forewarned is forearmed: If you have a vested interest in any aspect of business, politics, or culture, you've got to get a copy of Nicco Mele's new book—*The End of Big*—so you can actively manage the changes that are about to impact you the most."

—Joe Trippi, author of *The Revolution Will Not Be Televised*

"*The End of Big* covers the consequences of our constantly connected, technology-fueled society—and nobody is better qualified to write about it than Nicco Mele. This book is an honest assessment of the most complex and fast-moving parts of our world. Nicco keeps a watchful eye on the institutions we rely upon because they're now very much up for grabs. His continued focus is on protecting human values, our social structures, and our freedoms. Get a copy of this critical new book. It will forever change your thinking about business, politics, and culture."

—Howard Dean, six-term governor of Vermont

"Technology is redefining every aspect of existence—at work, at home, in the community, and in our private lives. Nicco Mele's *The End of Big* will help you chart a path forward that fits with your values, your world."   —Stew Friedman,
Practice Professor of Management,
the Wharton School, and author of
Total Leadership: Be a Better Leader,
Have a Richer Life

"We can give birth to a new kind of culture with institutions to match that doesn't cower at the technological advances but embraces that technology to bring us back to our communities so we can build a better future. For anyone who has ever felt overwhelmed by technology, take hold of this book. Now."

—Jeffrey Seglin, Harvard Kennedy School Lecturer

"If you'd like to better understand the rapid societal changes that technology has wrought, *The End of Big* is an indispensable guide. Nicco Mele provides an erudite yet supremely accessible look at how politics, media, business, and almost every facet of modern life has been transformed by the digital revolution—and prepares readers to make better choices and become more informed citizens."

—Dorie Clark, author of *Reinventing You:*
Define Your Brand, Imagine Your Future

"In *The End of Big*, Nicco Mele rightly argues that a now familiar tsunami of technology is changing our world in distinctly unfamiliar ways—with accelerating velocity. He also delineates a range of choices each one of us can make to reinvent business, politics, and culture for a better tomorrow, in light of his findings and insights. Read this cutting-edge book, understand your options, and start moving forward in new directions—for yourself, for your organization, and for the good of our collective future."   —Jeffrey F. Rayport, founder and chairman of Marketspace, and author of *Best Face Forward*

"Radical connectivity changes EVERYTHING, says Nicco Mele, and it's hard to disagree. From how we shop to how we work to how we govern, the end of BIG means the end of top-down, centralized hierarchical control. What it will look like when we get there remains blurry, but we can be assured that it will be radically different from the past. This book is an engaging guide to the underlying forces that are eroding all that is BIG, and its many examples will pull you in to this sweeping story of change."   —Amy C. Edmondson, Novartis Professor of Leadership and Management, Harvard Business School, and author of *Teaming: How Organizations Learn, Innovate, and Compete in the Knowledge Economy*

THE END OF BIG

NICCO MELE

THE

END

OF

BIG

HOW THE DIGITAL REVOLUTION MAKES DAVID THE NEW GOLIATH

PICADOR   ST. MARTIN'S PRESS   NEW YORK

www.picadorusa.com
www.twitter.com/picadorusa • www.facebook.com/picadorusa
picadorbookroom.tumblr.com

Picador® is a U.S. registered trademark and is used by St. Martin's Press
under license from Pan Books Limited.

For book club information, please visit www.facebook.com/picador bookclub
or e-mail marketing@picadorusa.com.

Designed by Anna Gorovoy

Permissions follow the index on page 331

The Library of Congress has cataloged the St. Martin's Press edition as follows:

Mele, Nicco.
    The end of big / Nicco Mele.—First edition.
        p.   cm.
    ISBN 978-1-250-02185-4 (hardcover)
    ISBN 978-1-250-02186-1 (e-book)
    1. Internet—Social aspects—United States.   2. Internet—Political
aspects—United States.   3. Information society—Social aspects—
United States.   4. Information society—Political aspects—United
States.   5. Computer network resources—Social aspects—United
States.   6. Computer network resources—Political aspects—
United States.   7. Power (Social sciences)—United States.   I. Title.
    HM851.M455 2013
    303.48'330973—dc23

                                                              2012037987

Picador ISBN 978-1-250-02223-3

Picador books may be purchased for educational, business, or promotional use. For
information on bulk purchases, please contact Macmillan Corporate and Premium Sales
Department at 1-800-221-7945, extension 5442, or write specialmarkets@macmillan.com.

First published in the United States by St. Martin's Press

First Picador Edition: May 2014

10   9   8   7   6   5   4   3   2   1

TO MY BOYS, ASA AND TOM—FOR YOUR FUTURE

CONTENTS

We shall not understand what a book is, and why a book has the value many persons have, and is even less replaceable than a person, if we forget how important to it is its body, the building that has been built to hold its lines of language safely together through many adventures and a long time. Words on a screen have visual qualities, to be sure, and these darkly limn their shape, but they have no materiality, they are only shadows, and when the light shifts they'll be gone. Off the screen they do not exist as words. They do not wait to be reseen, reread; they only wait to be remade, relit.

—William H. Gass
Gutenberg's Triumph: An Essay in Defense of the Book

THE END OF BIG

1

BURN IT ALL DOWN

*. . . "You've begun
to separate the dark from the dark."*[1]

Look around you.

Bloggers rather than established news outlets break news. Upstart candidates topple establishment politicians. Civilian insurgencies organized on Facebook challenge conventional militaries. Engaged citizens pull off policy reforms independent of government bureaucracies. Local musicians bypass record labels to become YouTube sensations. Twentysomething tech entrepreneurs working in their pajamas destabilize industry giants and become billionaires.

Radical connectivity—our breathtaking ability to send vast

amounts of data instantly, constantly, and globally[2]—has all but transformed politics, business, and culture, bringing about the upheaval of traditional, "big" institutions and the empowerment of upstarts and renegades. When a single crazy pastor in Florida can issue pronouncements that, thanks to the Internet, cause widespread rioting in Pakistan, you know something has shifted. When a lone saboteur can leak hundreds of thousands of secret government documents to the world, helping spark revolutions in several Middle Eastern countries, you again know something has shifted.

Technophiles celebrate innovations such as smartphones, the Internet, or social media as agents of progress; Luddites denounce them as harbingers of a new dark age. One thing is certain: Radical connectivity is toxic to conventional power structures. Today, before our eyes, the top-down nation-state model as we've known it is collapsing. Traditional sources of information like broadcast and print media are in decline. Aircraft carriers and other military hardware that for decades underpinned geopolitical power are obsolete and highly vulnerable, while organized violence remains a growing threat. Competitive hierarchies within industries are disappearing. Traditional cultural authorities are fading. Everything we depend on to preserve both social stability and cherished values, including the rule of law, civil liberties, and free markets, is coming unraveled.

The End of Big is at hand.

Institutions Aren't Dispensable

You might ask, Isn't the destruction of old institutions potentially a pretty good thing? Many traditional, big institutions are deeply flawed and even corrupt—they *deserve* to die. Few among us are not frustrated with the culture of influence and money in the two big political parties or disgusted by the behavior of at least one big corporation. Echoing the philosopher Oswald Spengler, isn't creative destruction, well, *creative?*

Our institutions have in fact failed us. Building a sustainable economy, for instance, that allows us to avert the catastrophic consequences of global warming seems hopeless in the face of big government, big business, and a dozen other big institutions. Ultimately, technological advances provide unprecedented opportunities for us to reshape our future for the better.

At the same time we can't fetishize technology and say "to hell with our institutions" without suffering terrible consequences. The State Department was designed and built for an era predating telephones and jet travel, let alone the distance-collapsing magic of the Internet. But that fact doesn't mean we should or can give up diplomacy. Government has become bloated and inefficient—but we still need somebody to repair roads, keep public order, and create the public sphere where the market cannot or should not dominate. Unless we exercise more deliberate choice over the design and use of technologies, we doom ourselves to a future inconsistent with the hard-won democratic values on which modern society is based: limited government, the rule of law, due process, free markets, and individual freedoms of religion, speech, press, and assembly. To the extent these values disappear, we're dooming ourselves to a chaotic, uncontrollable, and potentially even catastrophic future.

No, I Am Not Exaggerating

Ten years from now, we might well find ourselves living in constant fear of extremist groups and lone individuals who, thanks to technology, can disrupt society at will, shutting off power, threatening food supplies, creating mayhem in the streets, and impeding commercial activity. We've already seen a small group of hackers disrupt commuting in San Francisco for a few days (as a protest against police brutality) while flash mobs of approximately one hundred people have been showing up at retailers in cities like St. Paul, Las Vegas, and Washington, D.C., ransacking stores and making off with sacks of loot.[3]

This is just the beginning. Can you imagine daily life without currency issued by the national government? It's distinctly possible. What if in a hyperlocal society the sanitation department never comes to collect your trash—what would you do then? What if government bodies can no longer regulate the large numbers of small businesses that will exist with the End of Big? Could you trust that your food, medicines, and automobiles are safe? What will happen if authoritative news reporting ceases to exist and if cultural authorities fade into the background, inaugurating a new dark age? How will our democracy function? How will business advance? How will we solve big problems like hunger and global warming?

Wrapping Our Minds Around the Basic Problem

This book explores the destructive consequences of radical connectivity across many domains of contemporary society, from the press to political parties, from militaries to markets.

Other writers have examined the transformative potential of new technologies, generally focusing on specific domains such as business, economics, or culture, or on a specific dimension of technology's impact. This book seeks to address a broader problem that directly affects us all. Radical connectivity is altering the exercise of power faster than we can understand it. Most of us feel lost in the dust kicked up by the pace of change. We can tell political, social, and economic life is shifting, but we don't know what to make of it in the aggregate. Some changes seem destined to improve our lives, yet the impact on familiar institutions like the press makes us nervous. Opportunities for progress abound (and I will explore those, too), but so do openings for instability and even outright chaos. The devices and connectivity so essential to modern life put unprecedented power in the hands of every individual—a radical redistribution of power that our traditional institutions don't and perhaps can't understand.

Most of us, including policy experts, scholars, and politicians, haven't subjected radical connectivity to a deep and critical scrutiny, weighing the benefits and risks with a cold eye. Throughout the entire 2012 U.S. presidential primary campaign, not a single debate featured a substantial question about technology—about the nature and role of privacy for citizens, for instance, or about the disruptive impact of social media on the Middle East. But the problem runs deeper. We don't yet have an adequate vocabulary to talk about what's happening. The word "technology" is weak; a wheel is technology, and so is the printing press, whereas our present-day technology collapses time, distance, and other barriers. "Networked" doesn't quite capture the dramatic global reach, the persistent presence, the mobile nature of our world. You often hear "social" used in connection with technology—social media, social business, social sharing—but the consequences of radical connectivity on institutions are anything but social: They are disruptive, confusing, even dangerous.

Sometimes people utter the catchall term "digital," but it's not clear what that means, either; remember the digital watches of the 1980s? "Open" sounds good: open government, open-source politics, open-source policy. But WikiLeaks brings severe diplomatic and political consequences that "open" doesn't capture. Just because something is machine readable and online doesn't necessarily mean it is open. Also, "openness" describes the end result of technology, but it ignores the closed cabal of nerds (of which I'm one) that came up with this technology and defined its political implications. Not to mention that the control a handful of companies exert over our technology is far from open—companies like Apple, Google, and Facebook.

Ha Ha! You Laughed at Us, Now We Control You

This last point is critical—and as a prelude to this book, I'll spend the rest of this first chapter fleshing it out. Why has the digital age spawned so much social, political, and economic upheaval? Is it happenstance? Or is it a function of how specific groups of users have chosen to use technologies? Neither. A radical individualistic and antiestablishment ideology reminiscent of the 1960s is baked right into the technologies that underlie today's primary communications tools. Current consumer technologies are specifically designed to do two things: empower the individual at the expense of existing institutions, ancient social structures and traditions, and uphold the authority and privilege of the computer nerds.

Power is not about knowing how to use Twitter. It's about grasping the thinking *underneath* the actual technology—the values, mind-sets, worldviews, and arguments embedded in all those blinking gadgets and cool Web sites.[4] Without realizing it, citizens and elected leaders have abdicated control over our political and economic destinies to a small band of nerds who have decided, on our own, that upstarts and renegades should triumph over established power centers and have designed technology to achieve that outcome. "Cyber-activists are perceived to be the underdogs, flawed and annoying, perhaps, but standing up to overbearing power," says the tech pioneer Jaron Lanier. "I actually take seriously the idea that the Internet can make non-traditional techie actors powerful. Therefore, I am less sympathetic to hackers when they use their newfound power arrogantly and non-constructively."[5] Indeed, in our arrogance and optimism, we nerds haven't considered the impact of our designs, nor have we thought through the potential for chaos, destabilization, fascism, and other ills. We've simply subjected the world to our designs, leaving everyone else to live with the consequences, whether or not we like them.

Technology seems value-neutral, yet it isn't. It has its own worldview, one the rest of us adopt without consideration because of the convenience and fun of our communications devices. People worship Steve Jobs and his legendary leadership of Apple, and they consume Apple products such as the iPhone and iPad with delight and intensity—yet these products and indeed the vision of Jobs are reorganizing our world from top to bottom. The nerds who brought you today's three most dominant communications technologies—the personal computer, the Internet, and mobile phones—were in different ways self-consciously hostile to established authority and self-consciously alert to the vast promise and potential of individual human beings. The personal computer was born out of the counterculture of the late 1960s and early '70s, in part a reaction to the failure of institutions at that time. The Internet's commercialization took place in the context of the antiregulatory fervor of Newt Gingrich's Republican revolution and Bill Clinton's private-sector friendly, centrist administration. And mobile phone adoption exploded during the 2000s, challenging the institutions of the nation-state and bringing globalization to the digital world.

A Big Word: "Technopoly"

In his book *Technopoly*, the late cultural critic Neal Postman highlights the way the very technologies we design to serve us wind up controlling us. At first, Postman argues, we use technology as tools to help save labor and get the job done. Over time, the tools come to "play a central role in the thought world of the culture." Finally, in the technopoly, the tools no longer support the culture but dictate and shape the culture. "The culture seeks its authorization in technology, finds its satisfactions in technology, and takes its orders from technology." In reading Postman's work, I drop the word "culture" and substitute "big institutions." In our time, institutions like government, political parties, and the media seek authorization from technology and even take orders from technology.

The technopoly is not machine-controlled. There are people behind it, people with political ideals—as well as with economic and political interests—that they bring to technology design. The nerds whom mainstream society once portrayed as outcasts and undesirables are now powerful oligarchs, both literally and figuratively. McSweeney's *Internet Tendency*, an online literary magazine, ran a very fun piece titled "In Which I Fix My Girlfriend's Grandparents' WiFi and Am Hailed as a Conquering Hero."[6] If you can fix someone's technical problem, you suddenly receive enormous power and respect—a little bit of nerd ability goes a long way. Some of the largest, most powerful corporations on earth—Google, Facebook, Apple, Microsoft, Hewlett-Packard—are tech companies run by nerds. Joined by nonprofit organizations like Wikipedia, they shape our public life, our culture, and, increasingly, our institutions. Their products have reflected the political sensibilities of nerds both individually and collectively over the past fifty years. At its core, the technopoly's nonpartisan political philosophy can be summarized in one phrase reminiscent of the 1960s: Burn it all down . . . and make me some money! In parallel to this philosophy, with little sense of irony, the nerds in pursuit of market dominance have created a few new, really huge institutions of the digital age, the "even bigger" platforms we rely on every day, like Facebook, Twitter, Amazon, Google, and the lock-in universe of Apple's iPhone. These are the glaring exceptions to the End of Big that prove the rule.

Before later chapters of this book can examine radical connectivity's destructive effects on citizenship and community, political campaigning, news reporting, universities, scientific inquiry, entertainment, and corporate strategy, and before I can point the way toward the renewal of our institutions and a more stable and prosperous twenty-first century, we need to understand how a radical antiestablishment ethic became embedded in our technologies. Let me recount a few details of this well-worn story, the history of my people, the American nerds.[7]

Personal Computer Versus Institutional Computer

We take personal computers for granted, but in the history of computer science, they are a relatively recent detour. During the 1940s and '50s, most computer science took place inside large organizations—militaries, corporations, universities. Even by the late 1960s, the freshly minted computer nerd looking for a job would likely have gone into a large institution. That's because computers were giant institutional devices requiring a substantial amount of money and space. In 1969, Seymour Cray started selling the CDC 7600, a supercomputer whose base price was about $5 million. Imagine a wall of today's stainless steel side-by-side refrigerators, then build yourself a large office cubicle with walls made of giant stainless steel refrigerators. You've got the Cray 7600. Its top speed was about 36.4 MHz, not much compared to today's iPhone 4's 1 GHz (although both machines have the same flop score, a measure of how fast they can do floating point calculations). That means your iPhone 4 is about as powerful as a Cray supercomputer. Only large institutions could afford these bad boys, and they used them not to watch and share videos or listen to your neighbor's kid's noisy garage band but to perform complex calculations in mathematics, nuclear physics, and other disciplines.

Yet within this large, institutional world of computer science, a debate was erupting: Should we continue to build megacomputers that can essentially function as stand-alone, artificial intelligence machines, designed to tackle big, complex problems, or should we build small computers that augment and extend the capabilities of individual human beings? Most of the institutional energy through the 1960s and early '70s—not to mention the prestige and rewards—fell on the artificial intelligence side of things, while the small computer school was looked down on, their research compared with secretarial and administrative work. Which would you rather do—build a better administrative assistant or build a mind that could solve big problems? Today, small computing has won out—in part because of the momentum of Moore's Law.[8] While the work of building a big mind is still around (remember when IBM's Big Blue supercomputer beat Garry Kasparov at chess?), the small computer side has birthed a range of technologies culminating in your current Droid or iPhone. There is a particular hero to this strand of American nerd-ocity, one whose story begins to elucidate the political ideology behind the personal computer, and he is the computer scientist Douglas Engelbart.

"The Mother of All Demos"

A product of the greatest generation that fought World War II, Engelbart had a sense of the United States' grandeur and majesty when dedicated to a great challenge, and during the 1950s and '60s he was looking for the next great challenge. Inspired by Vannevar Bush's article "As We May Think," which championed the wider dissemination of knowledge as a national peacetime challenge, Engelbart imagined people sitting at "working stations"[9] and coming together in powerful ways thanks to modern computing. Using computers to connect people to build a more powerful computer, to "harness collective intellect",[10] became his life's mission. By early 1968, he showed off several of his inventions at a demonstration subsequently known as "The Mother of All Demos." Many facets of the modern personal computer were present at this demonstration in nascent form: the mouse, the keyboard, the monitor, hyperlinks, videoconferencing. Unfortunately, "The Mother of

All Demos" did not turn Engelbart into an instant celebrity out-side nerd circles. Even inside nerd circles, most of his colleagues regarded Engelbart as something of a crank. The idea that you could sit in front of a computer and actually work at it seemed lunatic in this age of massive institutional computers that worked for days to solve your complex problem while you did something else. You dropped a problem off with a computer and returned a few days later to find it solved; you didn't sit in front of it and wait.

Yet Engelbart's vision wasn't all that radical. Even as he imagined people sitting at computers and using them to aug-ment and extend their work, he still saw them as big, institu-tional things. In Engelbart's view, work stations would be thin terminals without much power—essentially shared screens plugged into one big computer. Personal computers as we think of them were not a part of his original vision, and in fact he resisted the personal computer revolution at first, working most of his life at big institutions like Xerox and Stanford. To get to the personal computer and the makings of the End of Big, we need to shift to a different strain of thought that was popping up at the same time in the nerd world, which received its most memorable expression in a book by another quixotic computer scientist, Ted Nelson.

Computer Lib

You've heard of "women's lib" coming out of the Vietnam era? Well, turns out there was "computer lib," too. Ted Nelson's pivotal 1974 book *Computer Lib: You Can and Must Understand Computers Now* confronted nerds everywhere with a rousing call to action, demanding that they claim computing for individuals so as to free them from the oppression of, you guessed it, large institutions. *Computer Lib* had a radical style similar to Stewart Brand's countercultural publication *The Whole Earth Catalog*, yet *Computer Lib* devoted itself to computers, offering both a primer on the basics of programming and a breathtaking vision of computing's future. The book's cover art—a raised fist, *à la* the Black Panthers—left little doubt about its intended radicalism. Computer science was burgeoning as a discipline at major universities. At the same time, much of the country was still caught up in the turmoil of antiwar protests and other social movements. Many young people were arguing that the government and the military-industrial complex had fundamentally betrayed the people by pursuing war in Southeast Asia. Students had gone so far as to organize protests against construction of large Defense Department supercomputers.[11] In this environment, Nelson's book landed like a tab of Alka-Seltzer in a tall, cold glass of water. It was provocative and clear in its anti-institutional leanings and influenced a generation of nerds.

Rereading *Computer Lib* today, I can't help but marvel at Nelson's prescience. In one section, he describes what it means to be online and in another he imagines a world of hyperlinks—decades before the Web was invented. One of my favorite quotes from the book is the following: "If you are interested in democracy and its future, you better understand computers." In 1974, the idea that thousands of people would have their own computers was not merely radical in a political sense—it was crazy talk. Computers cost millions of dollars. How could everyone have one? To his credit, Nelson also recognized the way nerds could form a new kind of institution that kept non-nerds at bay. In a portion of the book called "Down with Cybercrud," Nelson disparaged the half-truths and lies that nerds told non-nerds to keep them from understanding computers' power. He came out aggressively against the institutional nature of computers, hoping to bring them out of the big universities and military and into the homes of the masses, where they could serve what he saw as a truly liberating purpose.

Home-Brewed for the People

Inspired by Ted Nelson and others, a generation of nerds emerged from the late 1960s and '70s determined to disrupt the march of the institutional computer and bring the personal computer "to every desk in America," as Bill Gates famously put it. Brand described this generation as embodying a "hacker ethic": "Most of our generation scorned computers as the embodiment of centralized control. But a tiny contingent—later called 'hackers'—embraced computers and set about transforming them into tools of liberation."[12] This contingent went to work in their parents' garages and in their dorm rooms and eventually brought behemoths like Apple and Microsoft into existence. But the early products were considered hobbyist items on par with model trains and HAM radios, or, in today's world, DIY craft beers and handmade artisan soaps. One group, the People's Computer Company, put this explanation on the cover of its newsletter: "Computers are mostly used against people instead of for people; used to control people instead of to free them; Time to change all that—we need a . . . People's Computer Company."[13] It all amounted to a sharp departure from mainstream computer science in America, which lived on in the giant mainframes of academic and government institutions.

A famous example of the burgeoning anti-institutional computer counterculture is the Homebrew Computer Club, an ad hoc group of hobbyist nerds who in 1975 began meeting once a month in Gordon French's garage in Silicon Valley. Some of its more famous members included the Apple founders Steve Jobs and Steve Wozniak.[14] Gates drew the ire of the Homebrew Computer Club by selling something that had previously been given away free—a terrible development for hobbyists. Microsoft's first software product, Altair BASIC, was sold at a time that software was generally bundled with a hardware purchase. Homebrew members famously started to circulate illegal copies of the software at the group's meetings—arguably the first instance of pirating software. The Homebrew members were annoyed by a young Gates trying to sell software—it should be free! An angry Gates published an "Open Letter to Hobbyists" in the Homebrew Computer Club's newsletter, writing, "As the majority of Hobbyists must be aware, most of you steal your software."[15] Jobs and Wozniak, born of the Homebrew Computer Club, took a different approach. The ads that appeared in 1976 for their first Apple computer announced that "our philosophy is to provide software for our machines free or at minimal cost" and "yes folks, Apple BASIC is Free."[16]

1984

During the decade after *Computer Lib*, as personal computers became fixtures in American homes and as computer companies became established organizations in their own right, the notion that personal computers represented a naked challenge to the centralized power of both computing and larger institutions persisted. John Markoff's account of the counterculture's influence on personal computing relates how "[t]he old computing world was hierarchical and conservative. Years later, after the PC was an established reality, Ken Olson, the founder of minicomputer maker Digital Equipment Corporation, still . . . publicly asserted that there was no need for a home computer."[17] On the other hand, antiestablishment ideology became entrenched in manifold specifics of the PC's design; Markoff relates, for instance, that the visualization that comes with iTunes—the pretty colors that move and change in sequence with the music—was inspired in part by Jobs's use of LSD, which Jobs called "one of the two or three most important things he had done in his life." A liberationist ethic also became entrenched in the overt marketing of personal computing devices, most famously in a classic television commercial, Apple's 1984 spot.

Following the rousing success of Apple's first two home computer models, Steve Jobs wanted to do something big to roll out its third model, the Macintosh personal computer. He hired Ridley Scott, who two years earlier had directed the sci-fi classic *Bladerunner*, to make the commercial.[18] The result was a powerful and intense ad that referenced the dystopian future of George Orwell's classic novel *1984*. In the ad, a young woman breaks into a large auditorium where a crowd of mindless automatons sit listening to a giant screen of a speaking man, presumably Big Brother. The woman, representing the Macintosh (she has a sketch of the Mac on her tank top), smashes the screen. The advertisement closes with the text, "On January 24th, Apple Computer will introduce Macintosh. And you'll see why 1984 won't be like 1984.[19]

Recounting the story of the ad, industry journalist Adelia Cellini noted, "Apple wanted the Mac to symbolize the idea of empowerment, with the ad showcasing the Mac as a tool for combating conformity and asserting originality. What better way to do that than have a striking blond athlete take a sledgehammer to the face of that ultimate symbol of conformity, Big Brother?"[20] In many ways, this ad represents the apex of the personal computer revolution. The anti-institutional counterculture nerd ethos of the 1960s and '70s had grown into the personal computer industry—an industry large enough by this time that it could now buy an advertisement during the Super Bowl. Yet the advertisement it used to make a splash still paradoxically evoked a deep strain of radical individualism and personal expression in the midst of conformity.

The Internet Comes to Malaysia

Like the personal computer, the Internet reflected an "anti-big," individualist philosophy when it emerged during the mid- to late-1990s, albeit a strain of this philosophy flowing from the other side of the political spectrum. Not long after 1984, consumers of the personal computer wanted to connect these devices together so that they could share things. First, they wanted to share practical, physical things, such as a printer. By the mid-1990s, an Ethernet jack became standard issue in all personal computers. The arrival of Ethernet marked the Internet's true beginning, tacit admission that what was really powerful—and liberating—was not the personal computer but the person connected through the computer.

I remember how revolutionary the Internet felt when it first came out. During the mid-1980s, when I was in grade school, we had two computers—an early Apple IIe and an Atari ST. I loved that Atari ST almost as much as life itself, learning on that machine how to write code using Logo, a programming language designed for kids. A few years later, we got a modem, allowing our computer to talk to other computers over our landline. I immediately set to work figuring out how to use this technology. We were living in Malaysia at the time—my father was an American diplomat working for the United States Information Service—and Kuala Lumpur had a small but burgeoning hacker community. I first found them through bulletin board systems—BBSes as they were called—accessed through my modem. A BBS was modeled after a real, physical bulletin board; someone would volunteer a computer as the electronic bulletin board, and then other people could connect to that computer and post messages to the board.

One of my friends in high school hosted a BBS on his home computer. At night after his parents went to bed, he snuck around the house and unplugged the telephones so they wouldn't ring when people dialed into the BBS. Then he plugged the telephone line into his computer and waited for people to dial in. All night long people would call and connect to his computer, leaving messages on the electronic bulletin board. I, too, would stay up all night, dialing into different BBSes, getting into arguments about code and other subjects. Sometimes I'd get a busy signal and would have to wait to connect my modem all over again. Other times, a real person would answer the phone, irate that I had woken them at 2:00 A.M. When a new BBS came online, people in our little community got really excited. Once a new BBS with two separate phone lines came into service—wow, was that a big day. And yet, these BBSes were limited. I could share things with maybe twenty other nerds in my neighborhood, but I couldn't share things with a random nerd in another city, let alone another country. My computer was still its own little island.

I wish I could remember the day the Internet came to Malaysia, but I experienced it as a gradual evolution, like a tide coming in. For months, posts appeared on the BBSes about the coming of the Internet, although Malaysia as a country had not yet connected to it. In 1992, the government commercialized an Internet provider that had a satellite Internet connection from Kuala Lumpur to Stockton, California. It took a while for that provider to offer Internet access to individuals at home; initially only corporate or institutional access was available. Our family finally got online; I still remember we were subscriber number 117. The experience was electric. Suddenly, we weren't isolated any more. Living in Malaysia, I had to wait days or even weeks to get box scores for the New York Mets. I literally had to wait for a paper copy of the *New York Times* to arrive off a slow boat from China. Now I was connected to the Internet and could get those scores instantly, in real time! And not just the scores—the plays! Granted, I wasn't watching video, as I would today, but even the written recaps of games that I received and pored over amounted to a revelation, producing an incredible sense of joy and closeness to home. It felt like liberation, as if the rules of distance and administrative boundaries no longer applied.

A Merging of Acronyms

Like the personal computer, the Internet initially took shape within a big institution—the United States military. The Defense Department's Advanced Research Projects Agency (DARPA) had developed the idea for the Internet decades earlier, imagining a communications system that would facilitate the exchange of research from institutions scattered around the country. Over time, DARPA came to understand a second advantage to a network communications system: it could withstand a nuclear attack because it possessed no single point of failure. The Internet's physical infrastructure and design came together during the 1960s as part of an attempt to make communication between huge institutional computers easier. Bob Taylor, a DARPA computer scientist, had three monitors in his office, each connected to an institutional computer funded by DARPA (one each at UC Berkeley, MIT, and a private software company called SDC in Santa Monica, CA). Bob grew tired of having three different systems and wanted one way to communicate with all three. As he remembers, "For each of these three terminals, I had three different sets of user commands. So if I was talking online with someone at S.D.C. and I wanted to talk to someone I knew at Berkeley or M.I.T. about this, I had to get up from the S.D.C. terminal, go over and log in to the other terminal and get in touch with them. I said, 'Oh, man!' it's obvious what to do: If you have these three terminals, there ought to be one terminal that goes anywhere you want to go."[21]

It took a few years, but by early December 1969, four universities were connected to one another through a system called ARPANET: UCLA, Stanford, UC Santa Barbara, and the University of Utah. As time passed, more academic institutions joined the network. By 1981, 213 academic institutions had connected and it was becoming clear that ARPANET was primarily an educational network and not something that properly belonged under military purview.[22] By 1986, the National Science Foundation (NSF) was running the Internet's essential backbone, called NSFNET, and by 1990 ARPANET was shut down, although the military continued to manage a few key functions, such as the registration of domain names. During the late 1980s, the Internet also began to grow beyond the United States, connecting to educational institutions in Europe and Asia. In 1989, the first private company selling access to everyday customers in the United States opened its doors—a move that concerned scientists, who feared that the Internet would lose its research focus and become co-opted for other activities (online poker, anyone? porn?). In 1992, Congress got involved, passing a law encouraging the NSF to open the Internet to "additional users" beyond "research and education activities."

A Declaration of the Independence of Cyberspace

Here's where the story gets really interesting, from an End of Big perspective. By 1995, the NSF had relinquished control of the Internet's essential infrastructure to the Department of Commerce, removing the last restrictions on the Internet's ability to carry commercial traffic. When I asked people who participated how and why this happened, the overwhelming point of view was that legislators and regulators in Washington, D.C., simply weren't paying attention. They didn't understand the Internet's power; it had no place in their landscape of power, and no familiar analogue existed that would make it easy to grasp. Mitch Kapor noted in an April 2012 interview, "Nobody in Washington DC took [the Internet] seriously, so it was allowed to happen. By the time anybody noticed, it had already won."[23]

In effect, the Internet was released into the wild in a strong pro-business climate pushed by conservatives who wanted one big institution—government—to get out of free markets. The following year, Congress passed the Telecommunications Act of 1996, which deregulated the radio spectrum, allowing, among other things, the rise of huge media conglomerates like Clear Channel (paradoxical, I know). The underlying philosophy received memorable expression in a piece written by the technologist, cattle rancher, and Grateful Dead lyricist John Perry Barlow called "A Declaration of the Independence of Cyberspace." The piece begins, "Governments of the Industrial World, you weary giants of flesh and steel, I come from Cyberspace, the new home of Mind. On behalf of the future, I ask you of the past to leave us alone. You are not welcome among us. You have no sovereignty where we gather."[24]

Around the same time, Kapor published a front-page piece in the third issue of *Wired* magazine arguing that the Internet's architecture realized Thomas Jefferson's ideal of decentralization. The technological leaders of the time expected that the Internet's adoption would lead to positive social change and was better left untouched. In a recent interview with me, Kapor noted that among the technological pioneers of the era, "there was very little understanding of how human institutions work." The blistering Moore's Law pace of technology is simply not the pace of human institutions, and, even in the early 1990s, the technology had outstripped our institutions' ability to keep pace with it.

The nerd community's desire to see the Internet as a free and open space stems in part from the overbearing behavior of a single institution: AT&T. For years, AT&T's communication monopoly stymied the creativity of computer scientists and innovators, reaffirming in their minds the great distrust of large institutions that had taken root during the late 1960s. The man widely credited with inventing Ethernet and modern computer networking, Robert Metcalf, described it like this in *Vanity Fair*:

> Imagine a bearded grad student being handed a dozen AT&T executives, all in pin-striped suits and quite a bit older and cooler. And I'm giving them a tour. And when I say a tour,

they're standing behind me while I'm typing on one of these terminals. . . . And I turned around to look at these ten, twelve AT&T suits, and they were all laughing. And it was in that moment that AT&T became my *bête noire*, because I realized in that moment that these sons of bitches were rooting against me . . . To this day, I still cringe at the mention of AT&T. That's why my cell phone is a T-Mobile. The rest of my family uses AT&T, but I refuse.[25]

Barlow's radical libertarian ethic, hardened by the AT&T executives' contempt, layered itself on the earlier "burn it all down" ethic espoused by the personal computer's founders, helping to bring us the connectivity we know and love today: a free commercial space where individuals and companies can pursue their business largely (but not entirely) free of government's big institutional obstruction.

The "Openness" of the Internet

Not only was the Internet's propagation inherently individualistic; it was orchestrated to favor the flow of information between individuals and groups, unconstrained by organizations or hierarchies. The popular notion that "information wants to be free" dates from the 1930s, when academics in the engineering disciplines that gave rise to computer science started talking about the free communication of scientific knowledge. Over time, their philosophy morphed and took on an increasingly political definition. As with the personal computer, you can trace the technology and ethos that powers WikiLeaks as well as its philosophical underpinnings to the Vietnam era and the open-source movement then emerging from computer science. WikiLeaks is in part a reaction to the broader socio-economic dynamics of our time, but its challenge to the established power of the United States and its institutions is made possible by personal information technology and distributed open-source computing.

Today's communications technologies are also designed to foster the creation of new online groups capable of challenging established authorities. The Internet is not just a bunch of people; it's a place where anybody can start his or her own group without asking for somebody's permission. David Reed, who was one of the inventors of TCP/IP, the fundamental language of the Internet, is famous for postulating that the power of any network is exponentially related to the ability of the nodes of the network to form groups within that network. The best way to understand Reed's Law is to imagine a fax machine. To send a fax to a hundred people, you have to send a hundred faxes. To send an e-mail to a hundred people, you only have to send one e-mail. You have spontaneously formed a group—a group of a hundred people to send a single e-mail—within the network of the entire Internet.

Reed's Law suggests the enormous power unleashed when barriers to group forming fall. Today, anyone can form a group at any time for any reason and organize to grow the reach and impact of that group. The implications for existing institutions are evident most clearly in politics. When Barack Obama ran for president, his Web site let anyone—you!—start their own group as part of my.barackobama.com. I myself formed such a group and found that these groups made the network exponentially more powerful. Because I started my own group, I felt ownership and put time into recruiting and organizing members. Thousands of groups like mine played a pivotal role in Obama's ability to raise funds, secure the Democratic nomination over a competitor with strong establishment backing, and win the election. (By the way, the role of technology in electing Obama is yet another paradox: democratically decentralized groups in the service of a historically centralized institution, the president and the executive branch).[26]

Clarifications and Caveats

Even as I recount this history, I need to acknowledge that I am simplifying things. For instance, I am lumping all nerds together under one banner, whereas in reality nerd life and thinking cover a range of viewpoints clustered around two dominant ones. One view is the powerful notion that, as we've seen, institutions are not to be trusted. The other amounts to a somewhat contradictory quest for market dominance. Throughout the book, I contrast the power that technology pushes to the individual with the giant, nearly monopolistic platform companies that serve this power. These companies—familiar names like Apple, Google, Microsoft, Facebook—are fundamentally pragmatic businesses that see (and exploit) market opportunity, even when it threatens significant social institutions. Jeff Hammerbacher, formerly the manager of the Facebook Data Team and founder of Cloudera, has been often quoted as saying, "The best minds of my generation are thinking about how to make people click ads. That sucks." It sucks even more when it comes at the expense of institutions that have been critical to the success of Western-style democracy.

I also don't want to imply that the development of technology itself was entirely values-driven. Real linear innovation also helped steer technology toward empowerment of the individual. Moore's Law has effectively entered the cultural lexicon as shorthand for the speed with which technology naturally got small, fast, and cheap—and as such, conducive to the empowerment of individuals. It is hard—perhaps impossible—to separate cultural influences from the inherent technical drive toward smaller, faster, and cheaper, and, thankfully for our purposes, we don't have to. Multiple factors made us what we are today: a cultural and technological regime where the individual consumer really is king.

Too Flat for the Nation-State?

Over the past ten years, as personal, connected computers have become progressively smaller, the connections between them have gone wireless. Seven billion souls inhabit Earth, and there are over five billion active mobile phones—and counting. When you combine the personal computer, the Internet, and the mobile phone, you have the technical preconditions for our present radical connectivity. Anyone anywhere can reach anyone else anywhere at any time at practically zero cost—and share with them almost anything, from the poem they just wrote to a live video. The design and marketing of mobile technology does not contradict the anti-institutional ethic of personal computing and the Internet; on the contrary, it adds another layer, in which the big institution is not merely the United States federal government but nation-states everywhere.

You see the post- or antistate ethic at work in advertising for mobile phones, which often evokes the experience of having access to anything anywhere in the world, without regard for the national boundaries defined and administered by nation-states. Apple's ubiquitous iPod ads, for instance, simply show

a human figure enthralled and dancing to their music—
suspended in space, knowing no boundaries, experiencing total freedom. Innumerable other ads trumpet the global coverage
of mobile phones or the ability to sustain personal contact
with important people in your life across vast stretches of
geography and irrespective of national political boundaries.

Politically, such advertising jibes with free market capitalism and, specifically, with the pro-globalization politics espoused by many writers and policymakers on both sides of the
aisle, most famously by the *New York Times* columnist Thomas
Friedman. In his influential 2005 book *The World Is Flat,* Friedman argued that the world was shrinking and that many of the
traditional rules of economics were being rewritten as the barriers to competition were being flattened. Nation-states seeking to compete in the new global economy needed to get out of
the way and reduce administrative barriers that had conven-
tionally inhibited the flow of people, money, and ideas. Other
writers favoring free markets have gone so far as to proclaim
that the nation-state is effectively dead and meaningless as a
political entity in our age of globalization.

The nation-state may not be dead yet (it remains the basis
and means for all vital international agreements, from the World
Trade Organization to Interpol), but it is losing its cultural and
political force. "One of the huge changes brought by the printing press and advanced exponentially by the Internet is that
people are able to readily pursue different interests and points
of view," says Doug Casey, the best-selling author and chairman
of Casey Research. "As a result, they have less and less in common: living within the same political borders is no longer enough
to make them countrymen."[27] Professor Liza Hopkins agrees,
writing that "states continue to be an important container for

social identities, but the rise in global flows of bodies, goods and information now makes the nation only one of a number of competing sites of allegiance."[28]

Nerd Disease

As antiestablishment thinking has gotten baked into the technologies that assure radical connectivity, and as mobile technology has rendered its practical impact more pronounced, the influence of the nerds has only gotten stronger and more pervasive. The megaplatforms underpinning radical connectivity and created by the nerds constitute one glaring exception to the End of Big. The nerds also position themselves as a new "Even Bigger" institution in our age of eroding institutions by purposely building complexity into technology applications, thus creating a market need for their nerd expertise. I call this complexity "nerd disease," and I admit to benefiting from it myself: A substantial portion of my consulting practice involves explaining to CEOs that they need to combat nerd disease and start holding their technology and information officers more accountable.

Let me offer a quick but representative example of nerd disease in action. I once worked with a small company where the technology team was in one building across the parking lot from a warehouse that shipped product. The company, on the advice of its nerds on staff, was preparing to spend about $1 million (a large sum for this company) to integrate their e-commerce and warehouse systems so that when someone placed an order online it would automatically get sent to the warehouse for shipping. Early on in the project, as we started to figure out how we were going to pull this off technically, it became apparent that the project was growing increasingly complex, with commensurate increases in the cost estimates. The company's chief technology officer kept explaining to the nervous CEO that the integration of the systems was necessary, even with the increased complexity and cost. I finally suggested a different solution to the problem: Twice a day, have someone print out the online orders and walk them across the street to the warehouse. It would have taken several years of triple-digit growth before the cost of this simple solution would come anywhere near the cost of the complex but nerd-approved solution the company had been planning.

Nerds everywhere propogate their own authority by emphasizing complex, never-ending processes over outcome—a huge reason technical projects mushroom. Vivek Kundra, the first chief information officer of the U.S. government, was handed a stack of PDF documents representing $27 billion of IT projects that were years behind schedule and millions of dollars over budget. That's a classic case of nerd disease run amok—benefitting the nerds, enshrining them and their knowledge as a new authoritative institution. Kundra took an approach Justice Brandeis would have admired. He started a Web site called ITDashboard.gov and listed every project that was overbudget, how much it was overbudget, how long it was behind schedule, and a photo of the federal manager responsible for the project. Then he took a photo of the newly inaugurated President Obama scrolling through the Web site in the Oval Office. Despite pushback from the "IT Cabal," this measure of transparency created some accountability. Kundra noted the success of the project: "We were able to save $3 billion after terminating projects that were not working, or de-scoping them. We also halted $20 billion worth of financial systems last summer because these were projects we had spent billions on with very little to show to taxpayers."[29]

Trying to challenge the nerds at their own game is difficult at best, because the nerds also create their own myths of invincibility. A prime example is the Apple Store's so-called Genius Bar; apparently the nerds are geniuses who alone can understand radical connectivity and keep the whole giant ship running and heading in the right direction. That sort of pompous mythmaking is endemic. It's a shame, too, because most of our technology is at its core not especially complicated. Understanding its fundamental accessibility is the necessary first step toward taking real ownership of the technology and, with it, a new control over where our society is headed.

Looking Ahead

No single remedy available today will cure nerd disease, just as no single remedy will prevent the ongoing destruction of existing big institutions that remain vital to upholding social order and the Western values of democracy. As the next seven chapters argue, our leaders are on the verge of abandoning us to a chaotic world where the progressive gains of the last two centuries might be lost. Digital literacy is a critical solution, but perhaps most central is a deep awareness of our own values and a willingness to bring those values to bear in decision making. In the last chapter, I argue that the next decade will belong to those who can take the bottom-up, grassroots energy unleashed by radical connectivity and marry it with effective, engaged leadership to craft powerful political movements and build new businesses. We must make sure that we openly and collectively recommit ourselves to the values our nation was built on and to the building of new or updated institutions to replace old-style institutions in turmoil. Starting with the newly empowered, small entities, we have to discipline and inspire ourselves not merely to burn it all down but also to build it back up. Technology and smallness aren't just the problem confronting us—they are also the solution, the only real path to a safer, healthier, more prosperous future. It's up to us to walk this path purposefully and decisively, even in the face of understandable emotional responses like anxiety, uncertainty, and fear.

By way of introduction, this first chapter has offered a brief analysis of how a radical antiestablishment worldview became embedded in the technologies underlying radical connectivity. The challenge is to understand the worldview embedded in our technologies and then redesign institutions to bring technologies into alignment with limited government, the rule of law, due process, free markets, and individual freedoms of religion, speech, press, and assembly. We must marshal our best thinking, talent, and energy to imagine the institutions of the future; nothing less than the continued progress of the human race is at stake. It is my greatest hope that this book will inspire and engage people to build enduring institutions of the future, whether that means starting from scratch or reforming existing institutions to fit the reality of radical connectivity.

In the true spirit of the Internet, this book tends to go wide rather than deep. I try to paint a vast panorama that gives you what we so often lack: a sense of control over a truly unwieldy and confounding subject. As a result, I have not been able to dive exhaustively into the many subject areas I cover. Almost every chapter might occupy a single book on its own, and in many cases entire books have been written about material I try to address in a few pages. I have done my best to stay true to quantitative and qualitative assessments of the state of affairs in different disciplines, but those of you with domain expertise will undoubtedly take issue with certain of my points. I hope such mental sparring will not distract you from seeing the large picture, thinking about the way our world is changing, and helping craft solutions. In addition, not all chapters are created equal. The impact of the technology varies in its severity from chapter to chapter. Sometimes I scratch my head and struggle to figure out if the End of Big is in fact a problem in a given industry; other times, the End of Big seems manifestly catastrophic. This is in keeping with social and cultural phenomena generally, which tend to develop unevenly over time.

This book is ultimately about the nature of power in the digital age and how it overwhelms virtually every establishment it touches. Although we haven't yet realized it, our societies are on the cusp of a transformation as dramatic as the one the Athenians wrought when they decided to elect leaders instead of choosing them by birthright. We have a tremendous opportunity to reimagine the kind of society we want to live in and bring it into being; in this sense, I am far from a tech reactionary. But unless we understand our present and emerging technological regime, we will get nowhere, and we may even find ourselves at the mercy of a chaotic and unforgiving anarchy. *The End of Big* seeks to bring clarity and purpose to the digital lives we each live right now. It plots a course for the citizen of the present future, prompting us to grasp the moment that is already presented to us, so that we might lead each other toward a better, more humane, more progressive world.

2

BIG NEWS

"Look to your romantic interests and business invest-
ments," says the star hack in the newspapers. But
what if you have neither?
Millions will be up to nothing.[1]

Late on a Sunday night in May 2011, Keith Urbahn, Donald
Rumsfeld's former chief of staff, tweeted, "So I'm told by a rep-
utable person they have killed Osama Bin Laden. Hot damn."

Before any mainstream reporters got involved, this "tweet
heard around the world" was already spreading like wildfire.
"Within a minute," one observer notes, "more than 80 people
had already reposted the message, including the *New York Times*
reporter Brian Stelter. Within two minutes, over 300 reactions
to the original post were spreading through the network. These
numbers represent the people who either retweeted Keith's

original message or posted a reaction to it. The actual number of impressions (people who saw Keith's message in their stream but didn't repost it) is substantially higher."[2] Alongside this activity, a Pakistani IT consultant named Sohaib Athar, who lived near Osama Bin Laden, was unintentionally live-tweeting the raid, posting updates like "Helicopter hovering above Abbottabad at 1 A.M. (is a rare event)."[3] Around the world, many people first learned of Bin Laden's death and followed the ensuing story not by watching a network newscast or reading a newspaper front page but by reading posts circulated via Facebook, Twitter, and the like.

Consider four other recent world-changing events: the assassination of John F. Kennedy, Watergate, the fall of the Berlin Wall, and the tragic events of September 11. In each of these previous events, professional journalists working in the mainstream media played a crucial and even iconic role in delivering and distributing the news to a mass audience. In the case of President Kennedy's assassination, Walter Cronkite's broadcasts on CBS News came to stand as a shared cultural experience for millions of Americans. Who was not moved watching the great journalist choke up while announcing the president's death? The names of two respected newspaper journalists—Bob Woodward and Carl Bernstein—likewise became indelibly linked to the Watergate scandal. Network television footage of crowds at the Berlin Wall in the darkness of November 9, 1989, provided unforgettable, collective images of the end of the Cold War, echoing President Reagan's powerful rhetoric a few years earlier: "Mr. Gorbachev, tear down this wall!" And burned into the consciousness of many Americans following September 11, 2001, is the television footage of the airplanes plunging into the Twin Towers, not to mention the figure of Mayor Rudy Giuliani, standing at the podium answering questions at press conference after press conference with a mixture of calm, grief, and determination.

Nowhere has the threat to existing big institutions seemed so much in evidence as in the news world. Not long ago three big television networks and a handful of big newspapers domi-nated the news; now, thanks to the Internet, less than 40% of people get news from traditional sources,[4] with many turning instead to upstart blogs like *Talking Points Memo* and platforms like Twitter. Audiences have evaporated and journalists are be-ing displaced as "sources go direct";[5] in essence, large news me-dia companies are being scooped by their own sources—even by their own reporters on Twitter. A former senior cabinet offi-cial like Keith Urbahn probably knew a few big reporters who could have gotten news of Bin Laden's death out. Sohaib Athar, living in an urban part of Pakistan, might not have enjoyed the same reach, but a journalist might have conceivably tracked him down a few days after the event and asked his opinion as a local witness. Yet both bypassed traditional news reporting and found a global audience listening. Most big news opera-tions would need to wait for official confirmation or a second source, but on Twitter, you just go. Not surprisingly, Big News organizations are fast losing traditional sources of revenue, and in response they have changed their business models, stream-lined their news gathering, and in some cases closed their doors.

Should we feel glad that radical connectivity has rendered the news far more casual, scattered, deprofessionalized, and low-budget than ever? On the one hand, yes. By taking apart journalism's traditional business models, and by effectively democratizing news gathering, the demise of Big News allows us to remedy some of the recent ills of the news establishment, namely its increasing refusal or inability to hold power account-able. As Alex Jones has described it, traditional journalism's ability to serve as a watchdog comprises its "iron core"—its primary political function in a democracy.[6] We didn't see enough aggressive truth telling during the Second Gulf War or the period leading up to the 2007 financial crisis—indeed, we hardly saw any!—and the situation is worsening as news organizations shift resources away from investigative reporting.

Yet it remains unclear if emerging, upstart news organizations empowered by radical connectivity can ensure the kind of accountability Big Media did in its heyday. In the end, Dean Starkman may be right when he writes that, "journalism needs its own institutions for the simple reason that it reports on institutions much larger than itself."[7] Without Big News, we might well leave ourselves open to corruption and abuses of power the likes of which we have never seen. Unless we carefully build a watchdog capacity into today's emerging media institutions, the End of Big could see an infinite wash of user-generated "news" that fails to arrest the decay of our democratic institutions and the rise of corrupt and unscrupulous demagogues.

"Sources Go Direct"

Thirty years ago, even twenty years ago, a dispassionate observer might have reasonably judged Big News the wave of the future. The last half of the twentieth century saw a long historical trend toward media consolidation—the rise of increasingly big news-gathering organizations and institutions. Printing presses for newspapers were expensive, after all, and so were radio and television stations. To support capital investments, news organizations needed economies of scale and the advertising revenues generated by mass audiences. Big Business, meanwhile, needed Big News as a "reliable" vehicle for reaching their customer bases. I include quotation marks in that last sentence because as retailing tycoon John Wanamaker famously said, "I know that half of my advertising doesn't work. The problem is, I don't know which half." Marketing gun Seth Godin points out that it's actually closer to 1% of advertising, at most, that delivers the desired effect for companies.[8] Still, buying and selling a lot of advertising was a crucial part of corporate sales and marketing, generating personal fortunes in Big News and the advertising industry in the process.

Today, the six big companies that dominate the news industry—General Electric, Walt Disney, News Corp, TimeWarner, Viacom, and CBS—are in free fall, and Big Business buys a lot less advertising than before (we're not even talking about struggling Big Newspaper companies like Gannett, Tribune, Hearst, and the *New York Times* Company). In part, the reversal in fortunes owes to digital technology's helpfulness in figuring out which advertising works and which doesn't. But fundamentally it owes to the fragmentation of media—digital technology's dramatic redistribution of the audience to a huge variety of places, not to a handful of media outlets.

In 1980, you could buy a television ad during the evening news hour on one of the three networks—CBS, NBC, and ABC—and reach almost every single American home. Since then, the audiences for these networks have dwindled, dropping by more than 10 percent in the last decade.[9] During the 2000s, audiences moved away from cable news networks, too, and toward a profusion of sites on the Internet, while local television news remained stagnant. Once these audiences found themselves online, they started participating in the reporting alongside traditional news media. As the former *San Jose Mercury News* reporter Dan Gillmor noted in 2004, mainstream journalists needed to start thinking not about their audience but "their former audience." Matt Welch summarized the situation well for the *Wall Street Journal*: "What once looked like a permanent empire is now revealed as an ahistorical lucky stream undermined by overreach and the desire among captive citizens to be free."[10]

Free they can be, because printing presses (and radio stations and television stations) aren't as expensive as they were. Everyone with a smartphone carries around these capabilities all rolled into one, thanks to radical connectivity. And more and more, it feels like everyone is taking advantage and producing their own news—what is commonly called user-generated media. In 2004, Internet users globally could choose from just over 4 million blogs, with about 12,000 new blogs created every day.[11] In 2011, that number had risen to 164 million, with the top 12% claiming over 100,000 hits every month.[12] Meanwhile, more than 200 metropolitan newspapers have shut down during the past several years. Major cities like Detroit and New Orleans no longer have a daily newspaper. Since 2007, thirteen companies that publish daily newspapers have sought Chapter 11 bankruptcy protection,[13] among them the Tribune Company, which owns eight daily metro papers, including the *Los Angeles Times* and the *Chicago Tribune*; Star Tribune Holdings Corp., which owns the Minneapolis *Star Tribune*; the Philadelphia Newspapers, LLC, owner of the *Philadelphia Inquirer* and *Philadelphia Daily News*;[14] the Sun-Times Media Group, Inc., anchored by the *Chicago Sun-Times*;[15] and Lee Enterprises, owner of the *St. Louis Post-Dispatch*.[16]

In some cities, the newspapers that once played a major role in urban life now barely scratch the surface. In 2008, when Boston Mayor Tom Menino (the longest serving mayor in the city's history) found himself with the first significant challenger in two decades, the *Boston Globe* didn't even assign a reporter to the campaign full-time until a few weeks before the election. For months, the candidates duked it out, with local radio and a handful of local political blogs the only coverage.

Dave Winer calls the proliferation of blogs and other small, grassroots news and opinion Web sites "sources go direct," the "same idea" as user-generated media "but with more respect and emphasis on quality."[17] "Sources go direct" undercuts the economic model of news properties by fragmenting audiences and, more directly, through "the great unbundling." In June 2011, an FCC report on the condition of local news described the phenomen this way:

> During the news media's most profitable days, in many towns, there was only one newspaper, leaving consumers with limited choice. And, though we may not have thought of it this way, purchasing a paper meant having to buy a bundle of goods, even if readers only wanted certain parts. A cross-subsidy system had developed, in which a consumer who bought the paper for the box scores was helping to pay the salary of the city hall reporter.[18]

Similarly, William S. Paley (who built CBS) used to tell his reporters that the entertainment division paid for the news. These days, if all you want are the sports scores or the classifieds or the stock prices or even just the horoscope—well, you can get any one of those online without paying for the rest of the newspaper. Part of the collapse of newspaper advertising revenue owes to the great unbundling and the rise of a multitude of smaller news sites capable of fulfilling consumers' needs in a more tailored way.

Breaking Through in Rural Vermont

If you're an editor at a Big Newspaper, the End of Big might seem pretty grim. But if you're an underdog or an outsider, "sources go direct" can feel liberating, much like the personal computer did during the 1960s to people like Steve Jobs. Early in my career, I moved from Washington, D.C., where I was working in politics, to New York City in pursuit of a nonpolitical job. In New York, I felt cut off from the political gossip of the nation's capital—and, most importantly, from a publication called "The Hotline," put out every day by the *National Journal*. "The Hotline" chronicled every single political development in the country at the national and state levels. If you wanted to follow the Montana Senate race two years before the election, "The Hotline" was the place for you. But "The Hotline" cost a ton—as much as $5,000 a year during the late 1990s. As a young computer programmer, I couldn't afford a personal subscription.

Luckily for me, burgeoning online communities enabled me to indulge my obsession with politics. The Montana Senate race might get one or two articles in the *New York Times* in the race's final weeks. Being a political nerd, I desperately wanted to follow every twist and turn months or even years out. I discovered a rich ecosystem of political blogs run almost entirely by hobbyists, people with no connection to the political or media establishment. And they were following every race, every issue, in minute detail—trading tips, gossiping, arguing, and generally getting deep into American politics. It was a form of "sources go direct," and, without it, I would have found it impossible to follow the political fortunes of candidates across the country in a daily, coordinated, coherent way.

In April 2003, I left New York and moved to Vermont to work for Howard Dean's presidential primary campaign. At the time, most major media stories about the primary race started with a sentence that ran roughly like this: "John Kerry, Joe Lieberman, Dick Gephardt, and five other Democratic presidential candidates. . . ." My candidate, Howard Dean, was never mentioned by name. But hundreds of bloggers were writing about Howard Dean. Again, these blogs were run not by professional journalists but by hobbyists; they were grassroots, user-generated media. And so in response we started a blog for the presidential campaign—the first blog launched by a major presidential campaign[19]—to let "sources go direct." If Big News didn't care about Howard Dean and wouldn't write about him or cover his candidacy, then we'd cover it ourselves, taking our message directly to the public. Such user-generated media allowed an outsider candidate like Dean to bootstrap his presidential bid and gain real traction in terms of name recognition and fund-raising—something that Big News would never have allowed.

Big, BAD News

One reason, then, that Big News has come under siege is that it hasn't provided sufficient opportunity for underdogs and outsiders to break in. Yet Big News has also fueled its loss of market share by abdicating its traditional duties as an objective, impartial news-gathering system.

Over the last couple of decades, media organizations have catered more and more to the powers that be, concerning themselves too much with securing their continued, privileged access to sources and with turning news into a profit-making enterprise. It's a travesty, for instance, that during the last Gulf War, not a single major U.S. media outlet seriously scrutinized

the Bush administration's claims that Iraq possessed weapons
of mass destruction, not to mention the administration's as-
sertion that Saddam Hussein somehow helped orchestrate the
9/11 attacks. The *Washington Post* reporter Walter Pincus's
dogged reporting kept getting buried—including his infamous
story laying bare the Bush administration's claims that got
moved from the front page A1 to page A17. Kristina Borjessen
(formerly of CBS news) interviewed twenty-one major journalists about the press's failings around coverage of the Iraq War.
"The thing that I found really profound," she's said, "was that
there really was no consensus among this nation's top mes-
sengers about why we went to war. . . . [War is the] most extreme activity a nation can engage in, and if they weren't clear
about it, that means the public wasn't necessarily clear about
the real reasons. And I still don't think the American people
are clear about it."[20]

Less than a decade later, even as professional news gather-
ers began to admit their complicity in the nation's march to
war, the traditional press told the story of the greatest financial crisis since the Great Depression by essentially parroting
the talking points of Wall Street and the White House. The Pew
Research Center's Project for Excellence in Journalism reviewed
roughly 9,950 stories over the nine months in 2009 when the
financial crisis was unfolding, examining stories from major
news outlets across all media: television, radio, cable, newspapers, and online. Fully 76% of the economic stories studied
were reported from just two cities: New York (44%) or metro
Washington, D.C. (32%).[21] The Pew study concludes that the
mainstream media's coverage of the financial crisis was "told
primarily from the perspective of the Obama Administration
and big business."

As Wall Street veteran Max Wolff remarked, the media

"mostly did unpaid press releases for various businesses look-ing to sell financial products . . . becoming cheerleaders instead of critics and that took off the table, and out of the discussion, a critical voice that would have helped people realize what was going on, stop it before it got too big and deal with the crisis in a way that was relatively transparent, democratic and broadly beneficial—as opposed to being partial, muddy and unclear."[22] The *Columbia Journalism Review* has likewise documented how journalists covering the financial sector, even vaunted critics like Michael Lewis, routinely get paid outsized speaking fees by the very financial firms they are expected to cover with a critical eye.[23]

While the mainstream press fell asleep at the switch, "sources went direct" and started blogging. The Wall Streeter Susan Webber had already enjoyed a distinguished career in finance and management consulting when she started blogging under the pen name Yves Smith. Her blog Naked Capitalism did the kind of reporting she wasn't seeing, presenting titles like "The Beginning of the End?" in early 2007, well before the mainstream media was writing about the looming financial crisis. For instance, on the same day (Jan. 28, 2007) that "Naked Capitalism" ran a blog post titled "When Did Housing Lending Standards Become So, Umm, Lax?"[24] The *New York Times* ran a story about how cheap and plentiful home renovation loans were, titled "Hate the House? Then Renovate."[25]

The Rapidly Shrinking Newsroom

Big News's degeneration goes well beyond a couple of big misses in coverage. Keeping their eyes fixed on profit rather than pub-lic service, cable news organizations have cut back on outlays for investigative reporting, investing instead in cheaper opinion and talking heads. As Clay Johnson noted in his book *The Information Diet*, Fox News led the way, cutting back to 1,272 staff members in seventeen bureaus, less than half the resources its competitor CNN had maintained. MSNBC noticed Fox's rising profitability and declining reporting and followed suit, reducing its staff to only 600 people in four bureaus by 2010.[26] A similar story holds true for newspapers. In 2004, the *Los Angeles Times* garnered a net profit of more than $200 million, a profit margin of approximately 20%, and five Pulitzer Prizes for excellence in reporting. Corporate management pressed for cuts and greater profitability, and as a result the *Times* newsroom is today a shadow of its former self. Even though the LA Metro Transit Authority has a budget in the billions of tax-payer dollars and helps millions of Angelinos get to where they are going, the *Times* does not have a full-time reporter staffed to cover it.

The Incredible Shrinking News

It's tempting to indulge in schadenfreude and celebrate Big News's demise at the hands of upstarts and amateurs. But not so fast. The journalistic establishment existed for a reason: to ask tough questions, inform the citizenry, and hold power accountable. Who is going to do that in the age of radical connectivity?

Take presidential politics. Under President Obama, Whitehouse.gov has blossomed into a full-service news site featuring articles, video pieces, photo slide shows, and podcasts. The site's traffic has skyrocketed, to the point where it competes with MSNBC.com.[27] Although such a "sources go direct" approach can serve the public interest (FDR's fireside chats brought the country together even as they exerted pressure on FDR's congressional opponents), a withering independent media severely impedes the vetting process so central to democratic elections. It seems entirely possible that a candidate for president or an incumbent like Obama could make his or her way through a primary season without answering a single tough question. Relying on Facebook, Twitter, and friendly news outlets, major candidates can "go direct" and no longer need to engage the more objective or critically disposed mainstream media. As a result, citizens will have a harder and harder time finding objective information on which to base their decisions.

The demise of serious journalism is another primary concern. Bloggers breaking news direct from sources can increase public knowledge and oversight, but this practice doesn't replace the kind of dogged (and expensive) investigation that allowed the *Boston Globe* to break the Catholic clergy abuse scandal, or that allowed the *Washington Post* to bring to public scrutiny the explosive growth of the intelligence community. Traditional news, even in its weakened form, still accounts for much of the material we encounter in nontraditional media. As Alex Jones has noted, it provides fuel for everything from the opinion talk that dominates cable news to the comedy of late-night television talk show hosts. In 2010, a major study showed that blogs depended on traditional news media and on newspapers in particular, with more than 99% of blog links to new stories going to traditional Big News organizations. Between the great unbundling and the hunger for profit in large corporations, "demand will rule, and that may well mean that, as a nation, we will be losing a lot of news."[28] Throughout this book, we shall see examples of the same basic dynamic I'm describing here: small organizations continuing to piggyback on the same big institutions they are helping destroy.

Professional News Matters!

As a digital native and a nerd, I didn't always appreciate the social importance of traditional Big News, however imperfect these institutions may be. Today I do. In 2010, Alex Jones invited me to help judge the 2011 Goldsmith Awards, given by Harvard's Shorenstein Center on the Press, Politics and Public Policy (where I teach) to recognize exceptional investigative journalism. As part of my duties, I had to read hundreds of submitted articles that unveiled the worst failings of humanity, from abuse of power to accounts of serial killers to crippling corporate greed and more. The stories were so intense that I had to stop reading them at night. I came away inspired by the heroic work of many reporters who had dug into issues of enormous local, national, and global significance. And I was surprised to discover that the quality and depth of every submission transcended the vast majority of the output of bloggers or citizen journalists. Professionalism in journalism might not be sexy, it might seem old-fashioned, but it does matter.

Consider what happened in the small town of Bell, California. The town's mayor paid himself ever-larger salaries—over $800,000 a year—and engaged in the most crooked, crony capitalism, handing out other oversize salaries to his friends. Meanwhile, through a combination of lies and intimidation, he kept the largely poor, Hispanic city population in the dark, continuing to cut critical municipal services while giving himself and his buddies pay raises. After months of research and work, the *Los Angeles Times* delivered a pounding exposé of the corruption that led to multiple arrests. The town establishment had gone out of their way to stonewall various investigations, yet the *Times* had been able to break the story thanks to its financial and legal resources. Today, the former mayor of Bell is serving a ten-year prison sentence, a testament to the hard-hitting investigative journalism of Big News.

Most of the Goldsmith Award finalists that year were expensive, in-depth investigations of people with powerful and wealthy interests who were subverting the law and damaging our communities. In addition to the *Los Angeles Times* the finalists included the following:

- National Public Radio, whose deep dive into the powerful bail-bond industry found that it hurt defendants, their victims, and taxpayers. NPR's reporting has been cited in county commission meetings in Florida and in the statehouses in Virginia, Florida, and North Carolina.

- ProPublica, in collaboration with NPR's *Planet Money* and Chicago Public Radio's *This American Life,* who revealed how the 2008 recession was significantly deepened by the machinations of Merrill Lynch, Citibank, and Magnetar, a little-known hedge fund. As a result, the SEC is investigating deals referenced in the series.

- The *San Jose Mercury News,* which performed the first comprehensive examination of the influences of outside interests in California lawmaking. As a result of the investigation, legislative leaders proposed rule changes, and outside groups are pushing for mandatory disclosure of all meetings between lawmakers and lobbyists and greater disclosure of campaign contributions from sponsors.

- The *Washington Post,* whose "Top Secret America" described a massive expansion of government in the wake of 9/11. The two-year-long project resulted in congressional investigations, a review of all intelligence programs requested by the Defense Secretary, and the CIA's reduction of contract workforce.

- The *Las Vegas Sun,* whose exposé on hospital care in Las Vegas won the Goldsmith Prize. After a two-year investigation, including the review of 2.9 million records, the *Sun*'s reporting identified preventable infections and injuries taking place in Las Vegas hospitals, bringing the practices of the hospitals to light and ensuring they would be held accountable.

When I read about the cutting of newsroom staff and the shrinking number of jobs for journalists around the country, I shudder. A healthy and active fourth estate is essential to the functioning of our democracy. Without the prying eyes of organizations like the *Los Angeles Times,* expensive and politically charged stories that hold power accountable will not be written. As a result, corruption and abuse of power will persist and grow. Citizens will find themselves increasingly disempowered, oppressed, and manipulated by unscrupulous interest groups and demagogues working behind the scenes. The democratic institutions and trust in the rule of law that comprise the core of our civil society will be at risk.

Can't User-Generated Media Pick Up the Slack?

Not necessarily. The institutions of Big News in this country emerged over the course of more than a century, fostering a culture that rewards journalism at its best through institutions like the Goldsmith Prize. Radical connectivity, by contrast, does not come with these values built in. There is no Goldsmith Prize for Investigative Reporting by a Citizen Journalist. There is no "Sources Go Direct" Award for Courage and Honesty. On the contrary, we see an impetus, if anything, to turn journalistic writing into a commodity increasingly devoid of moral content. The *Huffington Post*, heralded as a successful online news business built on online advertising, uses a system called Blogsmith to manage and monitor publishing. Blogsmith tracks the amount of time a writer spends composing a piece and then compares it with the advertising revenue it generates once published. New outlets like Narrative Science have algorithms that "write" news stories, especially formulaic stories like reporting on sports events and financial news. Robots writing the news! As much as I love robots, I can't see one winning the Goldsmith Award any time soon. We need people—specifically, dedicated, professional reporters—to do that.

Steven Johnson has argued that journalism is evolving from a hierarchical system rooted in newspapers to a more distributed ecosystem with a multitude of players, including both professionals and amateurs. His view of online journalism as an immature ecosystem makes it clear how desperately we need to create some new institutions. As radical connectivity collapses many of the assumptions of the news industry—the economic and revenue models, not to mention the exclusivity of media distribution and delivery—we need to reorganize news gathering in ways that embrace the reality of radical connectivity while also inheriting the best values of the passing era of Big News. Imagining and building new organizations is challenging yet hardly impossible. Part of the solution may be to rely on smaller journalism outlets with fewer than twenty journalists but with mainstream reach and innovative business models. In recent years, a crop of online-only outlets have emerged that deliver political and government news that has otherwise been pruned from traditional media. They are staying in business while also doing their part to hold power accountable. Let's take a look.

Don't Mess with Texas

On November 3, 2009, venture capitalist John Thornton and journalist Evan Smith opened *Texas Tribune,* a nonprofit news site covering Texas politics and public affairs. Operating out of offices in downtown Austin, not far from the state capitol building, the *Tribune's* staff of just over a dozen full-time employees doesn't just post the usual items—feature stories, blogs, videos, and the like—that form the mainstay of political journalism. Instead, readers can also access a large collection of information, including tweets by government officials and databases that answer questions such as how much government workers make; how big departmental budgets in the state government are; and how much donors were contributing to politicians.

The *Tribune* itself makes money. The site's management took an approach they call "revenue promiscuity," squeezing precious dollars from many revenue sources simultaneously. As of winter 2012, some 3,000 members were contributing $50 or more annually. Corporate sponsors pick up the tab for events and certain content. Readers shell out additional funds to pay for premium content. And major donors—eagerly sought out by the *Tribune*—give gifts of $5,000 and up. Two years into the experiment, the *Tribune* has banked more than $10 million in revenue, enough to support a staff of thirty-one full- and part-timers. Although this newsroom is much smaller than that of other Texas news organizations, the *Tribune* still manages to serve a whopping 200,000 unique visitors each month, attracting them with the kind of political information that other newspapers, magazines, and Web sites are cutting back on.

To get in front of as many eyeballs as possible, the *Tribune* deployed an unorthodox, low-cost distribution strategy that was heavy on social media. The site's Facebook footprint included some 18,500 "likes" as of February 2012, while 22,500 people had subscribed to the *Tribune's* Twitter feed. Smith explained that the site focused on assembling strong content and waited for people to discover it on their own; with social media and the dynamics of viral sharing holding sway, "push is a much more effective way of reaching people than pull."[29] Continuing to build on its success, the *Tribune* shows that a new organization *can* carry the torch of accountability and journalistic integrity formerly carried by "big" institutions at their best.

CSI (Crowd-Sourced Investigations)

The *Texas Tribune* is an example of a new organization providing accountability journalism, but Britain's venerable *Guardian* newspaper (founded in 1821) offers a compelling glimpse of an old big institution experimenting and trying to find its way. In recent years, the Guardian has offered a free blogging platform to online opinion leaders, hosted chatrooms and discussion boards, and emerged as an early convert to podcasting. The Guardian was also one of the first news organizations in the world to turn its Web site into an open platform for Web developers and aspiring citizen journalists everywhere, launching the guardian.co.uk application programming interface (API), allowing Web developers globally to use a wide range of Guardian content in their apps.

Perhaps the Guardian's most interesting innovations involve crowd-sourced investigative reporting. In 2009, more than 2 million pages of expense reports from members of Parliament in the U.K. were made available to the public. The Guardian put all 2 million pages online and asked their readers to help them read the expense reports and find story leads. In the first 80 hours, more than 170,000 pages—almost 20% of the total volume—were reviewed, thanks to a visitor participation rate of 56%.[30]

The Guardian has gone on to take this approach time and time again, most notably with publicly released records from Sarah Palin's tenure as governor of Alaska and with the giant trove of diplomatic cables made available by WikiLeaks. On one occasion, Paul Lewis, a Guardian journalist, was investigating the suspicious death of a deportee on a plane from the U.K. to Angola. Lewis needed to find people who had been on the flight who might act as witnesses, so he tweeted from his account, starting a hashtag named after the victim, #jimmymubenga. Lewis received several responses, including this one—"I was also there on BA 77 and the man was begging for help and I now feel so guility that I did nothing."[31]

The Guardian's culture of hacks and hackers has brought journalists and nerds together to serve the public interest.

Here's the thing: It's working. The newspaper has a print circulation of about 280,000, far from the largest in Britain, but it stands as the fifth most visited news site globally, even though it maintains a fraction of the staff and budget of other competitors.

Yet Another Intriguing Model: Pro-Am Journalism

I first met Amanda Michel on the Dean campaign, where she worked as an online organizer. In 2007 and 2008, Amanda designed (with Jay Rosen of NYU) and ran the *Huffington Post*'s "Off the Bus" initiative. The program recruited 12,000 people to cover the 2008 presidential campaign, training many of them to act as journalists. "Off the Bus" developed the idea of "pro-am journalism," bridging the gap between amateur journalists and professionals.

What I love about "Off the Bus" is its combination of many of the values of traditional journalism with the reach and power of radical connectivity. Michel has spoken widely and written in the *Columbia Journalism Review* and elsewhere about the program's decision to create a culture based on the standards of traditional journalism yet also harnessing the power of digital technology. As she describes it,

> More professional journalists should take their offline skills— such as interviewing sources—online, and learn to build and manage networks of sources to produce accurate information. . . . For new media, the reverse is true. While they can quickly aggregate and grow the ranks of citizen journalists, they must take much more seriously the professional side of the equation— the reporting and editing and verification.[32]

"Off the Bus" succeeded in breaking news and providing new depth to coverage of the presidential race, but it also succeeded financially: "More than five million people read otb's coverage in October 2008, and our tab for sixteen months of nationwide collaborative journalism was just $250,000."[33] For the 2012 presidential race, Huffington Post invited a traditional journalist, the former Newsweek chief political correspondent Howard Fineman to lead "Off the Bus." Meanwhile, Michel has moved on, bringing pro-am journalism first to ProPublica, where she set up a "reporting network" that anyone can join. The organization's pitch to prospective citizen-journalists provides a sense of the emerging and exciting relationship between journalistic professionals and amateurs: "With a team of two dozen investigative journalists, ProPublica can dig deep, burrowing down into opaque company portfolios or massive government programs.... Making sense of our new—and ever changing—landscape requires piecing together your stories neighborhood-by-neighborhood and state-by-state. That's where you come in."[34]

1,000 True Fans

Pro-am journalism is not the sole completely new model for journalism now emerging. Another is the cultivation by bloggers of ardent micro-audiences. In March 2008, Wired magazine cofounder Kevin Kelly wrote a blog post arguing that in the digital era, artists can survive so long as they generate 1,000 true fans, each of whom might pay $100 a year:

If you have 1,000 fans that sums up to $100,000 per year, which minus some modest expenses, is a living for most folks. One thousand is a feasible number. You could count to 1,000. If you added one fan a day, it would take only three years. True Fanship is doable. Pleasing a True Fan is pleasurable, and invigorating.[35]

Kelly's post sparked much discussion, including significant rebuttals. But over the last few years, a "1,000 true fans" model of journalism has begun to emerge. Josh Marshall represents perhaps the most successful and famous example. He was writing for the *American Prospect* in 2000 when, inspired in part by the Florida recount of that year's presidential election, he started his blog, Talking Points Memo.[36] Soon, Marshall's blog was breaking news, notably a report on Senator Trent Lott's controversial comments praising Strom Thurmond's 1948 presidential run and another on the Bush administration's politically motivated dismissal of U.S. attorneys.

In 2004, Marshall wanted to go to New Hampshire to cover the presidential race. He posted his travel budget on his blog and asked for readers to help finance the visit. As he recounts, he got the idea from another blogger who had solicited donations from readers to help him purchase a laptop computer he had needed. "So I put up in November or so of 2003 an item saying, 'I want to cover the New Hampshire primary for the blog.' After 24 hours, $6 or $7,000 had come in. At the scale my finances were at the time, that was a very very big deal. I shut it off because I couldn't think of what I was going to do with $6 or $7,000—except for going out and buying a new car to drive up there."[37]

Inspired by the success of his New Hampshire fund-raiser, Marshall went on to raise money from the community to hire two full-time journalists:

I had the idea that if I could hire a couple of reporters to do something like I was doing, we could make even more use of all this information that was coming in. The fund-raiser for TPM Muckraker was successful beyond my expectations. We raised a little more than $100,000. This isn't in contributions of $5,000; this is people sending in $10, $25, maybe $50—the occasional $100 and $250. But certainly 90–95 percent was $50 and under. That basically gave me the money to build the site, rent an office and hire two reporters for a year. My expectation—that proved to be true—was that after a year, it would grow enough that we could sustain it through advertising.[38]

Since launching his blog in November 2000, Marshall has managed to build his 1,000 true fans into a serious business. Although still a small newsroom, TPM continues to do a range of hard-hitting reporting with a team of twenty full-time staff and a handful of contributors.[39]

Story of a River

Online communities also are developing some of the watchdog capabilities of traditional Big News organizations. Wikipedia, a collaboratively edited encyclopedia and the world's fifth most visited Web site, retains some of the hard-won functions of the fourth estate. While Wikipedians are fond of noting that, "Wikipedia does not have firm rules," Wikipedia culture applies two fundamental values across all Wikipedia entries: neutrality and verifiability.[40] Wikipedia professes to "strive for articles that document and explain the major points of view in a balanced and impartial manner," avoiding advocacy and debatable assertions. The site also embraces verifiable accuracy by requiring that authors include third-party references: "[Wikipedia] articles should be based on reliable, third-party, published sources with a reputation for fact-checking and accuracy." Wikipedia and the news media thus bear an interesting relationship: Wikipedia needs the news media to report on issues and verify breaking stories, while Wikipedia ends up acting as a curated collection of some of the best available journalism on a given topic.

To understand Wikipedia's journalistic value, look at one simmering regional issue, the question of whether a casino should be constructed on non-Indian land in Oregon's Columbia River Gorge area. Throughout the 2000s, the issue had occupied a place in local political discourse, rearing its head during the 2002 and 2006 governor's races. But even political news junkies might conceivably have felt somewhat in the dark. Big media spent some ink covering large campaign contributions related to the proposed casino as well as some of the key moments in the ongoing battle, but the leading newspapers, radio stations, or TV channels didn't provide a broadsweep historical account of the issue in all its evolving complexity. A concerned citizen who scanned scattered news coverage might well have wondered whether journalists assigned to the issue really got the bigger picture themselves.

Wikipedia rode in to the rescue. Over the past several years, a handful of amateur researchers on the site had been constantly updating and improving the article on the Columbia River, compiling in one place academic knowledge and journalistic coverage of relevant topics such as hydroelectric power, man-made pollution, and fisheries management. A Wikipedia volunteer, Pete Forsyth, recounted to me how he had painstakingly looked up old coverage of the casino controversy in hopes of allowing others to educate themselves; nowhere else would readers find all the research easily accessible in a single location. Traditional media validated the Wikipedia article by linking to it, citing it as a definitive overview.[41] With a ten-second search, citizens were now in a position to make smarter, more informed decisions about a local subject that mattered greatly to them.

Storify

Another online community, Storify, supports traditional journalism's news delivery and watchdog roles by offering a place to summarize and collect social media from around the Web to tell a coherent story. If you want to follow a news story on social media, it can be hard, given the dispersed nature of social media platforms. Storify essentially allows anyone, including professional journalists, to pull from a wide range of social media to tell a story.

Andy Carvin, who runs online communities for National Public Radio, started using Storify to collect social media while reporting on the shooting of Arizona Congresswoman Gabrielle Giffords. He "grabbed" a handful of social media mentions relevant to the story, starting with the last tweet from Congresswoman Giffords before she was shot and including tweets from other eyewitnesses as well as breaking news reports and relevant YouTube videos.[42] Carvin went on to organize a number of Storify narratives around the Arab Spring, telling the stories of political uprisings in Tunisia, Egypt, and Syria by curating social media from people on the ground who experienced it live. Josh Stearns of the media advocacy group Free Press likewise began using Storify to track, confirm, and verify reports of journalists being arrested at Occupy protests all over the

United States. During the Occupy Movement's heyday he documented more than sixty-nine journalists who had been arrested in twelve cities around the United States.[43]

Unfinished Business

As such journalistic experiments suggest, the dangers wrought by the End of Big are not necessarily insurmountable. Yet they are real. Writing in *The New Yorker* back in 1960, A. J. Liebling famously observed that "freedom of the press is guaranteed only to those who own one." Radical connectivity has turned that statement on its head. Today, when the titans of Big News have fallen, when anybody can publish anything for a global audience at virtually no cost, we may enjoy freedom of the press, but such freedom may no longer have the same value it once did. We don't know yet whether sufficient resources are in place to enable journalism to fulfill its historic role as guardian of the public interest. The current crop of journalistic experiments is exciting, but can small, grassroots newsrooms really exercise the same kind of accountability as the larger ones did? Are the new media outfits we've surveyed sustainable over the long term? Do we still need a public space with a shared focus on specific news stories? And these questions lead to another: What should we as a society do to sustain the kind of resource-intensive news reporting that holds the powerful responsible for their actions?

In the years ahead, the need for honest, high-quality, hard-hitting journalism will prove greater than ever. The citizens of Bell, California, learned the hard way that, even in a small town, power corrupts. Large, powerful organizations and interest groups may be dying, but they will persist for some time in various forms, and we will still need our Woodwards and Bernsteins (backed by *the Washington Post*) to speak truth to power. As the End of Big takes on steam, a vibrant fourth estate will also help us grapple with something new: the emerging complexity that threatens to overwhelm us and make democratic governance and perhaps even social order impossible. We need to keep the iron core of journalism vibrant and strong—and it's up to us to imagine and build the institutions that will do so. In every area of life, as reigning institutions tumble, they will take with them norms, rules, reach, and authority, rendering the workings of power increasingly difficult to discern. As the next chapter argues, this trend is already true in national politics, where the two major U.S. parties are increasingly unable to perform their traditional role of weeding out extreme and undesirable candidates for office. Who if not our objective, empowered journalists will help us bring a measure of order and understanding to the chaos? Who will ask the tough and elucidating questions on which all democratic governance, in the end, depends?

3

BIG POLITICAL PARTIES

I believe Icarus was not failing as he fell,
but just coming to the end of his triumph.[1]

Is radical connectivity launching America's two big political parties, the Democrats and Republicans, toward irrelevance and even their eventual demise? It's a legitimate question, and a brief historical example from the Democratic side suggests why.

As the 1984 presidential election season got under way, Democrats were in trouble. The front-runner for their party's nomination, Walter Mondale, didn't stir up much excitement, and he faced an enormously popular incumbent Republican president—Ronald Reagan—who was also an incredible communicator and former movie star. Mondale represented the consummate Democratic insider. A former U.S. senator and vice president under Jimmy Carter, he had spent much of his public life preparing to run for president, working his way up the Democratic Party food chain. Conventional wisdom judged him unbeatable in the primaries because he had locked up the Democratic Party's fund-raising machinery and most of the critical endorsements, leaving his competitors out in the cold. Still, he remained an uninspiring figure widely regarded as unable to mount a credible challenge to Reagan.

Mondale easily won the Iowa caucus. But then, to everyone's shock, he lost the New Hampshire primary to the charismatic and handsome Gary Hart, a young upstart U.S. senator from the American West. Now Democrats who had resigned themselves to Mondale as the nominee got excited. They began sending money to Hart in hopes he might defeat Mondale.

That didn't happen. This was 1984; people could only contribute to presidential candidates by writing and mailing paper checks.[2] Since Hart's campaign was not even six months old, it didn't have an office address listed in the yellow pages. Democrats all over the country wrote checks to Hart, put them in envelopes, and addressed them to Senator Gary Hart, Washington D.C. Hart's Senate office had to forward this flood of mail to the campaign office in Colorado. Somebody then needed to open these checks, endorse them, and deposit them, and the campaign had to wait two to four additional weeks for out-of-state checks to clear.[3] Hart literally couldn't get the money in the bank fast enough to spend it so as to challenge Mondale in subsequent primaries. The Hart campaign remained disorganized and underfunded compared to the well-organized, well-funded establishment campaign of Walter Mondale. Mondale locked up the Democratic nomination—before losing forty-nine states to Reagan in the 1984 election.

Now fast-forward to 2007. The Democratic Party again had an establishment front-runner—Hillary Clinton—who had spent her entire adult life in Democratic politics. Having served in the White House for eight years as first lady, she enjoyed one of the best political fund-raising operations ever seen, honed by decades of fund-raising for Democrats around the country. Although many Democrats harbored reservations about sending another Clinton to the White House (if Hillary won and then was reelected, that would mean twenty-eight years of either a Bush or a Clinton as chief executive), her charismatic upstart opponent, Barack Obama, was relatively unknown and unproven, not having served out a full single term as U.S. senator. More than half of the Congressional Black Caucus had endorsed Clinton over Obama.[4]

Unlike Gary Hart, Obama was able to use the Internet to build a fund-raising operation that bypassed the establishment. Once a crack had formed in the Clinton armor with Obama's win in the Iowa caucus, Democrats all over the country started to send money to Obama, and this time the technical infrastructure existed to mobilize those dollars on the ground. Unlike in 1984, the challenger to a major party's establishment candidate in 2008 was able to win not only the primary but also the general election.

Obama's win was just the beginning. Changes in the accessibility of information and the ability to quickly organize people to solve complex problems are leading to shifts in the distribution of political power. In the age of radical connectivity, the advantages of being big—of being an incumbent officially endorsed by the big parties—end up being liabilities. The speed with which outside challengers can maneuver unencumbered by the hierarchy and weight of traditional institutions leaves the political establishment dangerously exposed.

This is a good thing, for the political establishment in both major parties has become dangerously corrupt, undemocratic, and divisive in recent years, focused on raising money above all else. Yet in hastening the demise of the parties and empowering upstarts, radical connectivity also paves the way for a dangerous populism to take hold of our political system. We get exciting candidates like Barack Obama who can shake up the system but also extremist or fringe candidates who, if elected, could bring the whole house down. We've already started to see such candidates on the right with the rise of the Tea Party, but they're plausible on the left as well. To make radical connectivity work for the public good, we need to invent new political institutions that do what Big Parties used to do before the culture of money took hold: identify civic-minded leaders and wise policies that broadly serve the public interest.

Before Television

If that last sentence sounded just a little bit nostalgic about the olden days of Big Parties, that's because I am nostalgic—a little bit. For most of American history up through the mid-twentieth century, political parties served a vital function by cultivating and vetting candidates. Although inarguably elitist, the parties (and the old-boy systems that comprised them) made sure candidates for major office deserved to be leaders—that they possessed some essential mettle or fitness for office. Bad apples aside, most of party rank and file evinced a strong sense of morality and social responsibility born of a class-based mentality— quite a shift from what we see today. As the *New York Times* columnist David Brooks has observed:

Today's elite lacks the self-conscious leadership ethos that the racist, sexist and anti-Semitic old boys' network did possess. If you went to Groton a century ago, you knew you were privileged. You were taught how morally precarious privilege was and how much responsibility it entailed. You were housed in a spartan 6-foot-by-9-foot cubicle to prepare you for the rigors of leadership. . . . The best of the WASP elites had a stewardship mentality, that they were temporary caretakers of institutions that would span generations.[5]

Before the age of radio and television, it was pretty hard to see candidates up close, so the political parties and their grand political conventions functioned as a process that delivered trustworthy leaders and policies to America. Closing themselves off in smoke-filled rooms, political leaders argued the pros and cons of particular leaders and eventually made political deals that effectively vetted candidates.[6] Sure, this system was undemocratic and excluded many citizens who did not make up the nation's white, male, Anglo-Saxon establishment. Still, by today's low standards, it seems in general to have produced leaders who for the most part were basically competent, knowledgeable, civic-minded, and capable of taking a longer-term view. Comparing the old system to what we have now, the journalist Elizabeth Drew, a longtime observer of American politics, puts it this way: "The quality of the politicians has gone down. Now I'm not nostalgic for the old days, because in some ways it was too cozy, it was less open, there was much less opportunity for minorities and women. But you had more people who were grounded on issues, thought about national issues."[7]

By the 1960s, the system of backroom dealings dominated by influential party bosses had already begun to break down.

Progressive leaders had for decades tried to open up the nomination system through more transparent state primaries, but they had faced an uphill battle against the powerful patronage system that was the backbone of the political parties. As the historian Michael Beschloss has written, John F. Kennedy's election in 1960 represented an important turning point:

> John Kennedy in 1960 was very popular among the American electorate, but among party leaders and members of Congress, these were people who had served with Kennedy in Congress for 14 years. They knew that he didn't take his job in the House and Senate very seriously. He was very absentee. He did not have a particularly shining legislative record. So if 1960 had occurred under the old convention system, Kennedy would have had a very hard time getting the Democratic nomination because he would have been rejected by all those people who had worked with him in Washington. Instead, 1960 is one of the first years in which presidential primaries had a very large influence on the nominating process."[8]

The Kennedy aide Ted Sorensen tells a slightly different story about the importance of primaries. As he saw it, party leaders would not have picked Kennedy at a convention because they feared that a Catholic could not win in the Protestant South. Sorensen quotes Kennedy as saying, "Could you imagine me, having entered no primaries, trying to tell the leaders that being a Catholic was no handicap?"[9] Kennedy went on to compete and win in the 1960 West Virginia primary, defeating Hubert Humphrey and proving a Catholic could win in the South.

Following the 1960 presidential campaign, the nomination process steadily opened up, fueled by radio and television. In part as a result of the fractious 1968 Democratic Convention, the Democratic Party allowed ordinary voters to influence the delegate selection process and consequently the presidential nomination. The Republican Party soon followed suit. Over the next few political cycles, candidates from outside the establishment of both parties gained ground as the entire political system adjusted to new rules that reformed the backroom patronage system, as well as a new media environment where television reached into every American home. In 1976, the antiestablishment candidate Ronald Reagan challenged sitting President Gerald Ford; on the Democratic side, a relatively obscure governor, Jimmy Carter, locked up the Democratic nomination without the establishment. The former governor of New York and establishment heavyweight W. Averell Harriman reportedly responded to Carter's nomination by saying, "Jimmy Carter? How can that be? I don't even know Jimmy Carter, and as far as I know none of my friends know him, either."[10]

Money, Money, Everywhere

Since the 1970s, establishment forces in both parties have aggressively reasserted control over the nomination process. One thing in particular has made it exceedingly difficult for outsiders to run and win: money. Thanks to the centrality of television and radio advertising, the costs of running national and local political campaigns have skyrocketed. The business of presidential politics has also grown more expensive due to the front-loading of the primary schedule. As the early contests in places like Iowa and New Hampshire have become more pivotal, candidates have needed to come into the race with a great deal of financing; otherwise, they'll never even get out of the gate. These factors have altered the functioning of the Big

Parties. Long the mother's milk of American politics, money has become the main reason the parties exist.

The major parties' fund-raising infrastructure, with its emphasis on major donors and bundlers, has become a new gatekeeper to the presidential nomination.[11] Money saturates the system, corrupting the very idea of public service and creating a culture of influence and financial success that permeates Washington, D.C., making its suburbs among the country's most affluent. The idea of running for public service has evaporated, and instead we find a careerist mind-set. Elizabeth Drew, in her damning assessment of American political culture, *The Corruption of American Politics*, describes it this way: "Everybody's busy . . . purchasing access to the people with power . . . there is a permeation now of money in people's goals and decisions and ambitions for life that there really didn't used to be."[12]

As Dick Durbin, a longtime U.S. senator from Illinois, relates, "I think most Americans would be shocked—not surprised, but shocked—if they knew how much time a United States senator spends raising money, and how much time we spend talking about raising money, and thinking about raising money, and planning to raise money." It's true: members of Congress spend most of their time on the phone, dialing for dollars. The two party headquarters—the DNC and the RNC—are just a couple blocks away from the Capitol and feature large rooms filled with tables and telephones. Party leaders encourage lawmakers to spend hours at these telephones, calling through lists of potential donors. Representative Peter Defazio, a Democrat from Oregon, described the situation: "If you walked in there, you would say, 'Boy, this is about the worst looking, most abusive looking call center situation I've seen in my life. . . . These people don't have any workspace, the other person is virtually touching them."[13]

Politicians eye the plush K Street lobbying career (or the career as lobbyists in state capitals) that beckons at the end of a successful stint as an elected official. In his 2012 book *Republic, Lost: How Money Corrupts Congress—and a Plan to Stop It*, the Harvard professor Larry Lessig tells the story of the Mississippi senator John Stennis. In 1982, Stennis was serving as chairman of the Armed Services Committee when he was asked to host a fund-raiser with defense contractors. "Would that be proper?" he asked. Lessig writes: "Stennis was no choirboy. But his hesitation reflected an understanding that I doubt a majority of Congress today would recognize. There were limits—even just thirty years ago—that seem as antiquated today as the wigs our Framers wore while drafting the Constitution."[14] Gone are the days, it seems, of the "wise men" chronicled by Walter Isaacson and Evan Thomas, the story of six men who helped shape America's response to World War II and the Cold War. Dean Acheson, Charles E. Bohlen, W. Averell Harriman, George F. Kennan, Robert A. Lovett, and John J. McCloy weren't elected, nor were they diverse; they were all wealthy white men from the East Coast establishment. Yet they brought to their work a degree of selfless service unimaginable in today's political culture.

The presence of corporate money and the decline of civic duty among political leaders have had a number of additional consequences. As Washington has become more money obsessed, it has grown more narcissistic, passing laws that reward incumbency and that seek to lock challengers out. It has also grown more polarized. With the presence of interest groups emptying their deep pockets, real solutions from any political perspective dry up, and the Big Parties move to the extremes, bringing the citizenry with it. A recent academic study on the political parties notes that "Parties no longer compete to win elections by giving voters the policies voters want. Rather, as coalitions of intense policy demanders, they have their own agendas and aim to get voters to go along."[15] Americans have responded by growing frighteningly cynical. Which call would your member of Congress rather take—a phone call from a middle-class constituent hundreds of miles away or one from a former colleague turned wealthy lobbyist who just gave your nephew a salaried internship? As of this writing, the U.S. Congress, the Supreme Court, and the executive branch have the lowest approval ratings since political polling started as a profession.

If politics has become an impregnable fortress of wealth and power, our democracy is all the poorer for it. But is the fortress really impregnable? During the last ten years, the Internet has showed up to the party uninvited, tearing down the financial barriers of entry erected by the establishment. By enabling politicians to tap large numbers of small donors for support, radical connectivity has paved the way for candidates with unorthodox views and backgrounds to gain a public profile and, in some cases, take power on both the local and national levels. That process continues today, with ambiguous portents for our democracy. It all started with Howard Dean's meteoric rise and fall during the 2004 presidential primaries, which is where my personal story comes into the picture.

The Howard Dean Adword

During the 1990s, I had worked in Washington, D.C., building Web sites for organizations, including Common Cause, a campaign-finance reform group. I loved politics but found the role of money distasteful. It felt like a complete subversion of the Founding Fathers' notions of government of, by, and for the people. At the same time, I was becoming steeped in open-source programming and a range of heady ideas—crowd sourcing, distributed network power, the wisdom of crowds—embraced by the nerd godfathers discussed in chapter 1. Leaving Washington to pursue a nonpolitical job in New York, I immersed myself in the growing world of grassroots political blogs, which reminded me of the online communities I had encountered in high school.

After September 11, 2001, becoming concerned by the growing momentum toward war in Iraq, I began to spend more and more time surfing the emerging political blogosphere, lurking on sites like MyDD.com. And it was there that I first encountered the story of a nonpolitician, a medical doctor who was running for president of the United States—Governor Howard Dean. In March 2003, a friend of mine invited me to a Dean meetup at the Essex Lounge in Manhattan. I arrived fashionably late, only to find the venue packed, a line snaking out the door and down the block. I waited around for a while and then headed home. I was never going to get inside in time to hear Howard Dean. But my curiosity was piqued. The energy of the crowd outside the Essex Lounge was intoxicating, beyond anything I had encountered in D.C. This was real politics, and it felt good to actually believe in something.

The more I learned about Dean, the more fired up I got. I decided I would make a contribution to his presidential campaign, the first politician I had ever contributed to financially. But I could not find his Web site to make the contribution. It was not coming up in a Google search, and it wasn't anything easily guessable, like HowardDean.com, DeanForPresident.com, or even Dean04.com. I finally found it buried in the search results: DeanForAmerica.com. On the Web site, I could not find the donate button; turns out to donate you had to click a button inexplicably labeled "A Prescription for Change, Click Here." Instead of giving the Dean campaign a contribution, I decided to throw them a bone and make it a little easier to find the campaign Web site. I bought a Google AdWord so that anyone who searched for "Howard Dean" or "Dean for President" would see a little box pop up to the right of their search results that said, "The Official Web site of Gov. Howard Dean's presidential campaign: DeanForAmerica.com."

I cannot remember what kind of daily budget I put on the Google AdWord; it was ample, but not dramatically generous (I was living on a modest salary in New York City). I do know that I bought the ad and forgot about it until a couple weeks later, when my credit card bill arrived. I found a whopping charge from Google. Logging on, I saw that the search volume for keywords relating to "Howard Dean" was skyrocketing. The graph looked like a hockey stick, a slow initial start and then a sharp upward angle.

Sitting at my desk at work, I excitedly called the campaign. After waiting on hold for an eternity, I asked to speak to whoever was running the Web site. A woman named Zephyr Teachout came on the line. I told her I was running the Google ads. She said, "Oh, I wondered who was doing that." I explained that I could not afford to keep buying the ads, but that someone on the campaign should. "I think you can raise some money online this way to easily offset the cost of the ads."

Zephyr was unimpressed. "We're pretty busy. We don't have anyone here who knows how to do that. If you want to do that, you should move up here and do it."

I stared into the phone incredulously. She wants me to move to Burlington, Vermont? To volunteer on the campaign, running Google ads? Does she think I'm crazy? I politely concluded the conversation and hung up. About ten days later, I packed my car and moved to Vermont. Zephyr's words had stayed with me. I had long aspired to join a presidential campaign, and with the wars in Iraq and Afghanistan escalating, I felt a particular urgency to change the country's course. Inspired by the Jeffersonian ideal of grassroots politics I found on the blogs, and sensing that the fledgling political campaign of this Vermont governor offered some opportunity, I started driving.

The Crack of the Bat

I showed up at Dean's Burlington headquarters midday on a Saturday. Clearing off the end of a creaky desk, I set down my alien-shaped, flat-screen iMac. Within an hour, I was working with Teachout and two other Dean staffers, Bobby Clark and Matt Gross, on overhauling the Web site. They seemed stunned that I could make changes to the Web site without calling anyone. Before my arrival, they had been calling a Web company in another time zone for every single edit to the site. Soon, my hours on the campaign grew impossibly long as the Web site became the hub of the campaign's activity, requiring constant care and feeding.

I had the right combination of technical and design skills— jack-of-all-trades, master of none—to take the campaign's online activity up a notch. But I arrived in Vermont ignorant of online fund-raising, and I found myself mostly reacting to challenges and situations as they arose. The campaign manager, Joe Trippi, had been following the rise of MoveOn.org and was convinced that MoveOn had a secret sauce that we needed to acquire to raise money online. We carefully studied them, tracking their model of e-mail acquisition through online petitions, followed by well-crafted e-mails to the petition signers designed to convert them into political donors.

Trippi courted MoveOn aggressively, asking them for advice and help. The folks there were assiduously evenhanded; they did not want to create a perception that they favored one candidate over another. MoveOn eventually offered all the presidential campaigns help, and we gratefully took them up on it. Zack Exley, a MoveOn senior staffer, visited our campaign headquarters and helped us understand how to raise money online.

For the Dean campaign, e-mail acquisition became an obsession. At the campaign's height, Trippi would call me at all times of the day or night and demand to know the net new number of e-mail subscribers since his last call. A little icon on the Web site ticked up as the campaign continued to grow. In an explicit attempt to mimic MoveOn, we began posting various petitions. One of our first and most successful was an online petition in response to a withering attack from the Democratic Leadership Council (a conservative Democratic think tank) immediately following one of the first candidate debates in May 2003. The petition went viral, bringing thousands of new sign-ups to the campaign.

From a fund-raising standpoint, Howard Dean's compelling storyline had great potential. Dean was the outsider, the only Democratic presidential campaign not headquartered in Washington, D.C., and the only Democrat courageous enough to challenge Bush on issues like the war in Iraq and the sorry state of health care. But with the end of the campaign's first quarter approaching, we needed to convert the online excitement around Howard Dean into cold, hard cash. Otherwise, the mainstream media and political establishment would continue to ignore him.

Borrowing from traditional fund-raising, we created a baseball bat visual and announced our intention to raise more than
$500,000 in four days before the end of the quarter. Simply an-
nouncing our ongoing progress toward that goal was a thunder-
clap in political campaign fund-raising; traditional political
campaigns closely guarded their fund-raising totals over the
course of the campaign as an important part of playing politics.
We didn't know at first how well the Internet would deliver for
us, but it turned out we had aimed much too low. Soon we blew
past $500,000, and after some debate, we set an absolutely out-
landish second-quarter goal of $7 million. As crazy as this goal
was, our supporters understood that success or failure lay in
their hands. We couldn't court and count on major donors—just
the average American. Consequently, citizens poured energy
into reaching our goal and then justly felt like they had accomplished something by donating $50 or $100.

Dean's progress toward the goal became a riveting political
drama, in part because it dramatized the role of the Internet in
allowing an insurgent candidate to meet and then beat establishment candidates—like the front-runner, Senator John
Kerry—at the money primary. Describing the physical debut of
a red baseball bat at an August 2003 fund-raiser in New York
City, Professors Jennifer Stromer-Galley and Andrea Baker observe that "the symbolism would not be lost on the thousands
of blog readers and Dean supporters who had been 'swinging
the bat' to raise funds for Dean."[16] Donating online meant that
you were part of the most exciting story in Democratic politics
in decades; what could be more exciting than overcoming impossible odds to become the front-runner for the nomination
to be president of the United States? I later heard stories about
a wide range of Washington D.C. establishment figures—the
press, Republicans, and Democrats—all glued to their computer

monitors, hitting reload to see if the bat was going to break its cap. Donna Brazile, who managed former vice president Al Gore's 2000 presidential campaign, noted at the time, "Dean has reached one threshold with the establishment that no one expected him to reach: He's been able to raise money." The former Democratic House majority whip Tony Coelho described the reaction more bluntly: "The establishment, they are bitching about Dean, but they don't have an alternative. Because of Dean, we're on the cover of magazines. . . . Our establishment people aren't anywhere."[17]

The story of the bat to this day gives me goose bumps. In the end, we raised over $7 million and marked a watershed moment in American politics. As the only technical Web staffer, I updated the bat graphic myself, using a graphics editor program to manually fill up the bat pixel by pixel. Demands for more frequent bat updates came closer and closer together, and my sleep deprivation accelerated. Jim Brayton, a volunteer who soon joined the campaign as staff, came in for periods so that I could sleep, shower, and eat. The money coming in allowed us to staff up our Internet team (we had thirty-five people at its height at our headquarters), invest in new technologies, and open new offices. We managed to move beyond manual updates to the bat graphic, allowing future bats to be deployed without the round-the-clock hands-on maintenance required the first time around. The promise and opportunity of small-dollar online fund-raising resonated across the political establishment, fueling a new generation of tactics and strategies bringing together online and offline activity for maximum impact.

And then, of course, it all came crashing down. Despite the Dean campaign's meteoric fund-raising, we managed a distant third in Iowa. The national news media kept Howard's famous "scream" on constant rotation, making it hard to pull the campaign out of the nosedive. We soldiered on—all the way through Wisconsin—but on a sunny day in March, Howard Dean announced he was pulling out of the presidential race.

The Establishment Undone

Howard Dean's presidential campaign provided a blueprint for how a candidate might use the Internet to challenge the establishment. Key elements include a compelling narrative, an appeal to people's frustration with the growing insularity of the political establishment, a grassroots campaign ethos, and an embrace of the fundamental technologies and ideas that have fueled radical connectivity. Since 2004, candidates on both sides of the fence have embraced this approach. Barack Obama successfully challenged the apparently unassailable Hillary Clinton, while Ron Paul unsuccessfully challenged John McCain.[18] Like Dean, Paul managed to dramatically outraise all of the other Republican primary candidates, breaking the rules of the "invisible primary." His online success paved the way for his son's dramatic victory in a battle for the U.S. Senate seat in Kentucky.

Of the roughly thirty-five people on the Dean campaign's Internet team, several went on to found companies that have had a lasting impact on political fund-raising. One of those companies was my own, EchoDitto. Our first client was the U.S. Senate campaign of a little-known state senator from Illinois, a man who harbored big ambitions but whose name would appear to have doomed his electoral ambitions: Barack Hussein Obama. Another company that grew out of the Dean campaign—Blue State Digital—went on to provide the strategy and technology for Obama's 2008 presidential campaign before being acquired by the advertising and marketing conglomerate WPP in 2011. Other companies like Chapter Three, a Drupal consultancy that grew out of a Dean campaign open-source project called DeanSpace, and Advomatic, which has built hundreds of Drupal Web sites for political causes and candidates, have remained small but influential. The innovation and game-changing experience of the Dean campaign proved an irresistible opportunity for a generation of technology and political entrepreneurs.

And yet, we should ask ourselves: How much of an influence has the Internet really had on politics? While the Internet has changed the nature of political fund-raising, it hasn't had much impact on the reality of political persuasion. In most political campaigns I've worked on, you see the numbers move if you take a poll on Monday, run a week of television advertisements, and then take another poll. The Internet, to date, has no comparable impact on undecided voters. In 2008, Obama raised about a half billion dollars online and then spent almost all of it on television and radio ads—not Internet outreach—in swing states. In 2012, Obama spent more than $70 million in online advertising, but almost all of that went toward cultivating small-dollar donors, not toward persuading swing voters.[19]

The question of persuasion in political campaigns raises fundamental questions about the Internet as a communications medium. To some extent, political campaigns rely on television because political consultants are incentivized to go to TV (they generally pocket a percentage of the television ad buy). But campaigns also turn to television ads because those ads deliver, working in a reliable, measurable way, minimizing political risk. The Internet, and digital interactions generally, haven't offered the same reliable communications impact, and that's in part because digital is a less passive medium than television. As a television viewer, you get interrupted with an advertisement about a political candidate whether or not you care. In the online world, people pursue things they're interested in—and consequently during their time online they are much less persuadable. The communications challenge thus goes far beyond politics: How will any project, product, or brand that needs to persuade people to try it or buy it get traction in the digital age? It's an open question, and when you see competitive political campaigns shifting the bulk of their spending from television and radio to the Internet, you'll know someone has figured it out.

To date, radical connectivity has opened up the political system to grassroots movements by offering new ways of organizing and shaking the money tree, yet the Internet has so far not managed to bring back what we lost with the opening of the two Big Parties during the 1960s: a system for producing crops of generally decent, competent, civic-minded political leaders. In fact, radical connectivity is proving quite dangerous by pushing our political system to unprecedented levels of polarization—a problem that is worsening by the day.

Academic scholars like Alan Abramowitz have documented the polarization that has crept into American politics since the 1970s, causing our candidates and our political parties to become weirder and more extreme.[20] The Brookings scholar William Galston has asked important questions about whether we can regard a polarized party system as healthy, concluding in the negative:

> [T]he collapse of the postwar consensus—on containing communism as the centerpiece of international policy, on government

as the Keynesian manager of the economy, on culture as a sphere
of contestation that should remain outside of politics—entailed
the loss of shared assumptions. The consequence was the reverse of the "more reasonable discussion of public affairs."[21]

In other words, our politics has grown increasingly polarized at
the expense of good governing and public policy. By enabling
formerly marginal candidates with little stake in the status
quo to move to the forefront of debate, radical connectivity has
taken a slow slide towards polarization and intensified it. According to National Journal, 2010 and 2011 have been the most
polarized years in Congress since National Journal began chart-
ing this trend thirty years ago.[22]

Populists with Pitchforks

The most important example of this trend is visible on the far
right, with the stunning rise of the populist Tea Party. In 2010,
the Republican establishment woke up to find that their chosen
candidate in eight U.S. Senate primaries had lost to an insurgent, start-up candidate supported by the Tea Party. Alaska,
Colorado, Delaware, Florida, Kentucky, Nevada, Pennsylvania,
and Utah saw voters reject long-standing Republican party
leaders, including two sitting U.S. senators who had long been
regarded as conservative standard bearers, Alaska's Lisa
Murkowski and Utah's Bob Bennett. Such a shake-up was unprecedented in American politics, and it already is hampering governance and pushing our country to the brink of insolvency.
Take the August 2011 U.S. debt ceiling crisis. President Barack
Obama faced down House Speaker John Boehner around the

question of raising the debt ceiling. As the journalist Matt Bai recounted in the *New York Times*,[23] Boehner was handcuffed by the extremity of his own members, particularly the Tea Party and their resistance to any kind of compromise. Their failure to find common ground led to the U.S. government's credit rating getting downgraded. As NBC News reported: "Economic activity stalled, and economic confidence tanked."[24]

The longtime Indiana senator Richard Lugar summarized the situation well following his loss to a Tea Party insurgent in the 2012 Republican primary. In his concession statement, Lugar wrote:

Too often bipartisanship is equated with centrism or deal-cutting. Bipartisanship is not the opposite of principle. One can be very conservative or very liberal and still have a bipartisan mind-set. Such a mind-set acknowledges that the other party is also patriotic and may have some good ideas. It acknowledges that national unity is important, and that aggressive partisanship deepens cynicism, sharpens political vendettas and depletes the national reserve of good will that is critical to our survival in hard times . . . Our political system is losing its ability to even explore alternatives. Voters will be electing a slate of inflexible positions rather than a leader. I hope that as a nation we aspire to more than that. I hope we will demand judgment from our leaders.[25]

We have reached a critical juncture. If we sit back and do nothing, the two Big Parties will continue to lose influence, and the public good will suffer at the hands of small groups with rabid, fringe views. We also run the risk of electing to public office extremist candidates who lack the basic knowledge required to perform competently in office. Although we have yet to see such candidates emerge in a serious way on the Democratic side, the 2012 election cycle saw a number of scary candidates emerge as front-runners among the Republicans, including:

- Minnesota Congresswoman Michele Bachmann, who confused New Hampshire with Massachusetts as the location of the start of the American Revolutionary War[26] and who went on to suggest a conspiracy among Democratic Presidents to start flu epidemics. In her words: "I find it interesting that it was back in the nineteen-seventies that the swine flu broke out then under another Democrat president, Jimmy Carter. And I'm not blaming this on President Obama—I just think it's an interesting coincidence." In fact, as the *New Yorker*'s Ryan Lizza points out, the first swine-flu outbreak occurred under the watch of Republican Gerald Ford.[27]

- Herman Cain, former business executive and longtime political activist, revealed a fundamental misunderstanding of the mechanics of the presidency's role in constitutional amendments and suggested that China's attempt to develop nuclear weapons was a crucial foreign policy priority—even though China has had nuclear weapons since 1964. Although a potential leader of the free world, Cain nevertheless felt free to declare: "I'm ready for the 'gotcha' questions and they're already starting to come. And when they ask me who is the president of Ubeki-beki-beki-beki-stan-stan I'm going to say, you know, I don't know. Do you know?"

- Texas Governor Rick Perry, who suggested on his first day campaigning for president that Federal Reserve Chairman Ben Bernanke may be a traitor to his country and raised the prospect of Texas seceding from the Union. An infamous moment came during one of the Republican presidential candidate debates when Perry forgot one of the three federal agencies he was campaigning to eliminate: "I will tell you: It's three agencies of government, when I get there, that are gone: Commerce, Education and the—what's the third one there? Let's see. . . . OK. So Commerce, Education and the— . . . The third agency of government I would—I would do away with the Education, the . . . Commerce and—let's see—I can't. The third one, I can't. Sorry. Oops."[28]

I am by no means contending that the Republican Party and the conservative movement enjoy an unusual capacity to attract unqualified freaks. Republicans count among them a number of compelling leaders, but these serious men and women—Indiana Governor Mitch Daniels, Maine Senator Olympia Snowe, and Nebraska Senator and Vietnam-era war hero Chuck Hagel come to mind—are unfortunately bowing out of public life and elected office. Following Senator Lugar's defeat, Hagel gave a long interview lamenting the departure of a generation of GOP leadership who had put national interests above party politics: "They made it work because their obligation and responsibility was to a higher cause than their party. They were all partisan but they all knew their higher responsibility was to move this country forward and resolve issues through compromise and consensus. We've lost that glue in the Congress."[29]

ActBlue

Even a serious politician like President Obama finds himself vulnerable to the "populists with pitchforks" phenomenon wrought by radical connectivity. If six million people can give $100 to elect you president, then six million people will give $100 to impeach you if they don't believe you're doing the job you promised to do. We have yet to feel radical connectivity's full destabilizing effects, but in the years ahead we just might. The question arises: Do remedies exist that might forestall mounting instability? Can the two Big Parties continue to exist with the End of Big? How do we save and nurture the democratic process?

We need a radical revision of our parties—or at least some new institutions to work alongside the old ones. Glancing about the political landscape, we can discern some promising initiatives that might help our current Big Parties adapt to the conditions of radical connectivity, so that they can temper the culture of money and serve once again as a farm system for vetting and championing leaders and policies in the public interest.

Shortly after the Dean campaign ended, two graduate students—one from MIT, and one from Caltech—founded ActBlue.com, a Web service that creates an online fund-raising page for every Democrat running for office in the entire country, whether or not they want one, and then allows people to build their own lists of favored candidates. ActBlue doesn't discriminate; as long as you're a candidate from the Democratic Party, they'll provide online fund-raising infrastructure for you. You can go online right now, create an account, and build a list of your favorite Democrats running for office. Voilà—you're a political fund-raiser! Now you've just got to find some folks to contribute to your list of favorite candidates.

ActBlue goes toe-to-toe with the very purpose of the Democratic Party—raising money—and in this way challenges the establishment. In the traditional political ecosystem, groups of individuals with shared policy outcomes band together, primarily through professional associations, and raise money to support candidates interested in the same policy outcomes. So the American Medical Association raises money from doctors to support legislation that affects doctors. But what if you're interested in something more obscure, or something without an organizing constituency? I want to reform copyright law—an unpopular stance in the Democratic Party because of the influence of major Hollywood donors. I can create a page on ActBlue and then add Democratic candidates from around the country whose positions on copyright reform I support. Now it's up to me to raise money to support them, tapping social media and my personal network for small contributions. It costs me nothing, and it costs the candidates on the other end nothing. If you're able to mobilize a constituency, you can use the free tools of ActBlue to build a small-dollar fund-raising network that can propel your candidacy or your issue forward, all while remaining well aligned with the Democratic Party.

Significantly, ActBlue is not itself part of the party. It is a curious entity, a nonprofit political committee that functions as the grassroots fund-raising backbone of the party. Soon after its 2004 launch, ActBlue became the Democrats' single largest source of cash, raising more than $200 million for Democratic candidates. What's amazing is that ActBlue is astonishingly civic-minded (as long as you're a Democrat). With the volume of cash flowing through the site, ActBlue could have acted as power brokers in the Democratic Party or converted themselves into a highly profitable private company. Instead, ActBlue's leaders have stayed true to their vision: "To democratize power by putting powerful fund-raising tools in the hands of Democratic candidates, voters, fund-raisers and donors across the United States."[30] Explaining why in an April 2012 interview, Adrian Arroyo, Director of Communications, noted that incorporation into the institution of the Democratic Party would threaten ActBlue's neutrality, which has been so vital to its success. "If we were a part of the Party, at some point there would be a intra-party schism, or a major primary challenge, and our infrastructure would come under institutional pressure to favor one side or the other. At this point we've got the necessary separation to be totally neutral."

ActBlue, then, can keep constituents and candidates aligned with the established Big Party, avoiding a Tea Party–type rupture, because it remains marginally outside the party structure. Refraining from taking sides when Democrats divide over issues, it exists in parallel to the party in order to save it. No counterpart to ActBlue exists on the Republican side, even though many Republicans seem to feel that their party needs one. What would happen if the United States government—perhaps through the Federal Election Commission—were to build an infrastructure similar to ActBlue, letting any American raise money for his or her favorite candidates?

ActBlue evokes a key paradox of the End of Big. It demolishes the "bigness" of the old institution—the "invisible primary" of money and major donor access—by taking advantage of a fundamental characteristic of radical connectivity: the ability to connect everyone equally. With enough time and energy, anyone can use ActBlue to build a political money power base within the Democratic Party—not just W. Averell Harriman or Hillary Clinton. And yet as a central clearinghouse, ActBlue is in its own way Big; it just hasn't wielded its big power yet, preferring to remain a boring (but vital) piece of infrastructure. To that extent, ActBlue may provide one important structure for revamping our parties so that they are democratic and yet still capable of producing serious leaders addressing serious issues.

Meetup

Of course, politics is about more than money. A functioning democracy requires a certain level of civic engagement—a subject of some concern among political scientists. As the Harvard professor Robert Putnam has famously observed, Americans have seen a generational decline in civic engagement. Over a few decades, Americans went from bowling in organized bowling leagues to bowling by themselves, even as more people started bowling. Putnam showed how over the same time span, Americans purchased more air conditioners, started watching more television, and built fewer homes with front porches. Overall, Americans were staying inside more, not interacting with the neighbors, and becoming more distant from their communities. And as they were becoming less socially engaged, they were also voting less and in general becoming more removed from the political process.

Inspired by Putnam's book *Bowling Alone*, two young Internet entrepreneurs in the wake of 9/11 designed a Web service that let people find affinity groups in their neighborhood. People would be able to go to Meetup.com, type in their zip code and an interest, and find out if a nearby group existed dedicated to that interest. If one doesn't, the site invites you to start one. I initially encountered Meetup at a Barnes and Noble in Clarendon, Virginia, near where I was then living. Thirty to forty people were sitting in the store and knitting. "Oh, that's the knitting meetup," the clerk told me when I inquired. "They're here the second Tuesday of every month." I have since attended a wide range of meetups: new puppy meetups, Westie meetups, reading meetups, learning Spanish meetups, new dad meetups, chess meetups, and more.

Meetup quickly took off, and it continues to grow at an impressive clip. Somewhere along the line, it became a crucial vehicle for American politics, validating Putnam's theory that social engagement corresponds linearly to political engagement. The first major political meetups occurred around Howard Dean's candidacy for president. In many ways, the Dean meetups formed the backbone of the campaign, a grassroots infrastructure that fueled much of our fund-raising success. People who thought they lived in conservative Republican neighborhoods would go on Meetup.com and discover a group of Deaniacs meeting right near their house. By the campaign's conclusion, more than 140,000 people were attending more than 600 Dean meetups in every part of the country.[31] Meetup attendees turned out to be incredible online contributors, since the face-to-face interaction created an intense bond with the campaign. As Dean explained to *Wired* magazine, "[Meetup attendees] built our organization for us before we had an organization."[32]

During the Dean campaign's early days, we entertained ourselves by competing for the largest meetup slot with two other groups, pug owners and witches. My campaign coworker Zephyr Teachout was obsessed with beating the witches meetup: "Witches had 15,000 members, and we had 3,000. I wanted first place."[33] Today, Meetup has more than 100,000 meetups worldwide with more than 15 million members. While Dean remains Meetup's most successful political candidate to date, Meetup has served as infrastructure for other political movements. As of this writing, just shy of 700 Tea Party groups exist on Meetup, with almost 100,000 members. Scott Heiferman has this to say about the direction Meetup has taken: "When we were designing the site, we were wrong about almost everything we thought people would want to use it for. I thought it would be a niche lifestyle venture, perhaps for fan clubs. I had no idea that people would form new types of P.T.A.'s, chambers of commerce or health support groups. And we weren't thinking that anyone would want to meet about politics, but there are thousands of these Meetups."[34]

Unlike ActBlue, Meetup is entirely neutral and unaffiliated (both the Occupy Wall Street and Tea Party movements use it as an important part of their infrastructure). Because of its loose, grassroots nature, Meetup might encourage political movements outside the Big Parties, but it can and does just as easily serve as a parallel structure that keeps more mainstream Republicans and Democrats engaged and affiliated with their parties. Since each Meetup leader funds his or her own group, it is difficult to use the platform to do anything big without grassroots support. The Republican party cannot just say, "Here are our meetups"; instead, individuals within the party would have to step forward and say, "I'm hosting a meetup in my community, and I'm going to organize and pay for it." Meetup can offer an intriguing way for citizens to experience grassroots engagement that maintains their alignment with the Big Party, even as they achieve some autonomy from it—an ability to organize outside the thrall of Big Money and establishment politicians.

Americans Elect

But what about entirely new alternatives to the Big Parties? AmericansElect.org was an intriguing experiment during the 2012 presidential campaign, a nonpartisan organization and online nominating process existing outside both the major parties and designed to put a presidential ticket on the ballot. The organization spent close to $22 million, mostly raised from a handful of major donors, many of whom remain anonymous, and managed to navigate the complex legal labyrinth to get on the 2012 presidential election ballot in twenty-eight states.[35] The group had hoped to run an online, Web-based convention over several weeks to select a presidential ticket, attracting some major candidates with national credibility. As the group announced: "Any constitutionally eligible, qualified citizen—no matter their party—can seek the nomination or be drafted by Americans Elect delegates. We will never promote any candidate of our own—and the winning ticket is chosen directly by American voters through a secure, online process. To ensure the integrity of the process, candidates will be verified and certified by an independent, nonpartisan committee and meet a set of standard qualification criteria such as background checks."[36]

Some observers have critiqued Americans Elect for its anonymous major donors and its ability to disrupt the presidential election by promoting a spoiler on one or the other side. The organization failed to select a candidate during its primary process, for no candidate had garnered enough support to participate in the organization's online convention. Still, Americans Elect stands as an exciting experiment, an attempt to forge an entirely new institution in American politics. Because of the way ballot law works, even if Americans Elect has played a small role in 2012, its investment in ballot access could reap dividends in future elections. The organization could conceivably grow in strength and become a significant political player in future election cycles, influencing the shape of the two Big Parties and perhaps generating additional, competing institutions that help our political process function better.

In some places, the revolution in personal choice brought about by radical connectivity is also busy creating a new kind of politics. Consider the Pirate Party that has sprung up almost out of nowhere in several European countries. The core concerns of the Pirate Party represent a shift away from a traditional left-right dichotomy, with a relatively narrow focus on digital issues including using technology to facilitate direct democracy, protecting individual privacy, substantially reforming copyright laws, and generally advocating something resembling civil libertarianism. It's easy to imagine a platform like the Pirate Party's having some success in the United States by sidestepping the ideology that generally splits Republicans and Democrats. To date, the Pirate Party has elected forty-five members of Parliament in Germany. In Sweden, the Pirate Party has enjoyed limited electoral success but claims to be the third largest political party in terms of membership. Nicholas Kulish, Berlin bureau chief for the *New York Times*, describes the Pirate Party this way: "These are the people who read those long privacy notices the rest of us guiltily click 'O.K.' on and try to forget about as we post intimate photos of family and friends on a Web site intended to earn a profit for a corporation. For them, politics, with its thousand-page pieces of legislation, is really the fine print of the social contract."[37]

No More Armored Hubcaps

Outside Western Europe, the rest of the world is witnessing manifold challenges to the political establishment. The Arab Spring (see chapter 6) is one example, but just as interesting is a growing global movement that demands more transparency and accountability from leaders, whether or not they are democratically elected. Global Voices (an amazing international community of bloggers), in partnership with Transparency International (a global coalition to fight corruption), has the Web site Technology for Transparency Network, which tracks more than sixty initatives in more than thirty countries outside North American and Western Europe,[38] all of which use radical connectivity to challenge the political establishment.

The *New Yorker* profiled one such initiative, Rospil.ru, in Russia, which crowd sources evidence of corruption in the Russian government. Russians submit evidence—photos of empty fields where there are supposed to be apartment buildings, government documents available online due to a new freedom of information policy—and an online community then examines them, rating them and assessing their validity. "The projects have ranged from strange data systems for the Russian military to a new, overpriced Web site for the Bolshoi Theatre. Most recently, [Rospil.ru founder Alexei] Navalny highlighted the request for an Audi 8L, armored to the hubcaps, for the finance minister of the Russian Republic of Dagestan, at a cost of three hundred thousand dollars. 'I'm positive that the presidents of many of the world's countries get around in more modest cars,' Navalny wrote. Five hours after the post went up, the request was cancelled."[39] Navalny, who found himself at the forefront of protests against President Vladimir Putin, has been called "arguably the only major opposition figure to emerge in Russia in the past five years"[40] and "the man Vladimir Putin fears most"[41]—in part because of the way he has harnessed radical connectivity to challenge the political establishment.

Chile has a particularly active transparency community, with impressive sites like that of La Fundación Ciudadano

Inteligente (the Intelligent Citizen Foundation). Ciudadano Inteligente makes it easy to track elected officials' disclosures of conflict of interest and compare those conflicts to their legislative portfolios. The Foundation is credited with changing the culture of money in politics in Chile.[42] A more direct approach is happening in India at a Web site called IPaidABribe.com. If you're in India and you're asked to pay a bribe to, say, get a birth certificate for your child or get your tax refund, you can go online (or even submit from your phone) an anonymous report of bribery. Bhaskar Rao is a major government official in Karnataka, one of India's largest states. He approached the founders of IPaidABribe.com and used data they provided to reform part of his administration. In an interview in the *New York Times*, Rao said, "It was my unofficial spokesman to drive home the message that the public was really upset about this corruption. . . . It helped me get my colleagues to fall in line, and it helped me persuade my superiors that we needed to do this."[43]

The Western world has its own transparency initiatives; in the United States, the Sunlight Foundation has done an incredible job at innovating technology to open up politics and government. As the Supreme Court Justice Louis Brandeis famously said, "If the broad light of day could be let in upon men's actions, it would purify them as the sun disinfects."[44] The Sunlight Foundation has launched a range of projects, from tools to better engage with your elected officials to an "influence explorer" that lets you look up a single company across several databases, so that you can see how much money it has contributed to political campaigns side-by-side with its lobbying expenditures, its federal contracts, advisory committees its executives serve on, and complaints filed against it.[45]

Toward a Better Kind of Leader

In a 2010 piece for the *New York Times*, the journalist Matt Bai said this about our present political era: "The most prevalent ideology of the era seems to be not liberalism nor conservatism so much as anti-incumbency, a reflexive distrust of whoever has power and a constant rallying cry for systemic reform."[46] By bringing previously marginal candidates to the fore, the End of Big has injected new life into parties that had long ago become corrupted in their pursuit of cash. Yet for every mainstream candidate like Barack Obama enabled by radical connectivity, we can point to poor parodies of leadership whose ascent to power seems scary indeed to contemplate, no matter what your political leanings. The traditional political parties favor candidates who are effective fund-raisers or personally wealthy; at its worst, the corrupted institutions of the current system combined with the disruptive power of radical connectivity yield unrepresentative candidates who thrive on a distracting kind of populism but remain unserious and unprepared for the challenges of leading our country. We need future leaders who are effective fund-raisers and populist communicators, but who are also much more: thoughtful, ethical decision makers who consider the impact of their policies on future generations; inspirational leaders who can rally our best efforts as a society; deft diplomats who can navigate the complicated, interrelated global world in which we live.

We must find a way to update our political system, imagining and building new institutions, either within the existing parties, in parallel to them, or utterly outside them. Some of the structures I've described here are a start, but they are only just that; we do not yet know if five or ten years from now we will find ourselves with leadership not merely incapable of handling or preventing national crises but also haplessly or intentionally creating new ones of their own. The old way of nominating leaders isn't working, and so it's time for all of us, regardless of party, to get involved and get creative. Together, let's build new institutions consistent both with our common American values and the realities of radical connectivity. The End of Big demands nothing less.

4

BIG FUN

I am in need of music that would flow
Over my fretful, feeling fingertips,
Over my bitter-tainted, trembling lips . . .[1]

Ever heard of Shaycarl? He was eking out a living as a granite and real estate salesman when in 2007 he got a $500 Dell laptop and discovered YouTube. "I could do that," he said to himself. He became a member of YouTube's "Partner" program in February 2008 and got his first AdSense check, representing a portion of advertising royalties YouTube earned from his content, the following April.[2] Today, Shaycarl and his family—BabyTard, PrincessTard, SonTard, RockTard, Katilette—have a successful online reality TV show about their quirky family life that is approaching half a billion views on YouTube and earning an estimated several hundred thousand dollars a year. No talent agents were involved, no Hollywood studios, no television channels; in fact, nothing big was involved—except for YouTube.[3]

Beyond its impact on the news media and politics, radical connectivity has reduced the cost of creating entertainment. Today anyone can create just about any media, and anyone else can copy or share that media at zero cost. On YouTube, forty-eight hours of video are uploaded every minute, or nearly eight years of content every day.[4] Just about any cheap laptop and video camera can produce high quality video suitable for broadcast television. My mobile phone even captures high-definition video. As Francis Ford Coppola observed in 2007, "Cinema is escaping being controlled by the financier, and that's a wonderful thing. You don't have to go hat-in-hand to some film distributor and say, 'Please will you let me make a movie?'"[5] The same is true of other entertainment forms, such as music and books.

Is the democratization of media production and distribution an unmitigated "wonderful thing"? Just as radical connectivity has threatened the business model of Big News companies, so, too, is it hurting the big companies built around entertainment: the six big book publishers, the six big movie studios, the four big television channels, and the five big music labels. Technology has changed audience sizes (more options mean smaller and smaller audiences), and created alternatives for creative talent to "go direct." On top of all these changes, it's a lot easier to obtain creative content without paying for it. Just ask Hy Strachman. Strachman is in his nineties, lives on Long Island, and for almost the last decade has pirated hundreds of top Hollywood movies, creating packages of DVDs to send to U.S. military soldiers stationed in Iraq and Afghanistan.[6] Without the knowledge or consent of the big corporations who own the rights to the movies he copies, he has churned out hundreds of thousands of bootlegged box office hits. If a bored ninety-two-year-old, five-foot-five World War II veteran can subvert the big movie studios from the comfort of his tiny apartment, costing them tens of millions of dollars—well, then the traditional business models underpinning commercial entertainment are obsolete.

Given how cheap it now is to produce and distribute media, the compelling profit margins enjoyed by the giant media companies cannot persist. That's right—there simply won't be as much money to be made by big companies. But there will be money to be made by small companies and by creators like ShayCarl with niche audiences. Will that be enough? Will people continue to make high-quality cultural products that require extensive capital and that don't lend themselves to the conditions of connective technology? How will artists and consumers fare when a few even bigger technology platforms like YouTube present us with the vast proportion of our commercially available fun?

From Napster to Radiohead

The End of Big arrived first in the music industry in the form of digital file sharing. Napster came on the scene in June 1999 and boasted over 60 million users at its peak some two years later.[7] But it wasn't just Napster. Ted Cohen, a music industry executive at EMI Music at the time, recalls, "There was a perfect storm around 1999 when manufacturers started putting CD burners in computers, blank discs went down to under a buck, and MP3 came along."[8] By 2009, total revenue from recording sales and licensing was $6.3 billion—cut in half in just ten years (from $14.6 billion in 1999).[9]

Although the big record companies have seen their business evaporate, people continue to make and sell music. When you step back and look at industry data, you might even see progress. Quantitative evidence suggests that neither quantity nor quality of recorded music has declined since Napster.[10] Indeed, more people are creating music on more platforms for more people than ever before. Rather than the big media companies, the changing economics of the music industry have tended to benefit small, independent labels. They don't have the overhead of big labels and don't need blockbuster hits to make their business model work. One academic paper notes, "independent music labels, which operate with lower break-even thresholds, are playing an increased role in bringing new works to market."[11]

The big label companies didn't help themselves much. Not only did they resist music's transition to digital (perhaps sensible from an economic point of view), but they also continued to look for new ways to exploit performing artists. Even as the recording industry went to war with Napster, the big record companies lobbied Congress to pass a law that would have prevented musicians from ever accessing the copyrights to their own music by reclassifying music recordings as "works made for hire." This effort made a number of musicians very, very angry—just at a time when the big recording companies needed artists on their side. The law was passed and later repealed, but the damage was done. Don Henley of the Eagles accused the record companies of talking out of both sides of their mouth, of trying to protect artist copyright from Napster theft while conspiring to steal artist copyright.[12]

These days, musicians continue to experiment with business

models. The band Radiohead, raised and nurtured inside a big recording industry, struck out on its own once its contract with a big label expired. The band frontman Thom Yorke was quoted at the time as saying, "I like the people at our record company, but the time is at hand when you have to ask why anyone needs one. And, yes, it probably would give us some perverse pleasure to say 'Fuck you' to this decaying business model."[13] Radiohead went on to make their next album available on their Web site for download, and it invited fans to suggest how much they were willing to pay for it. You could pay nothing, or you could pay $1,000—whatever you felt it was worth. While Radiohead has not released formal numbers on the album's performance, by all accounts the album made more for them than their previous record contract.

Grandpa Goes Global

It's not just musicians who are experimenting with "going direct." My grandfather, Pete Davidson, just turned ninety. During World War II, he served in the army and got stationed across the Pacific. He dutifully wrote letters home to his cousin Warren, and Warren kept all those letters. My grandfather still has all the letters he wrote home to Warren, and in many cases he has the photographs he took to go with them. This collection represents a treasure trove of family history and a firsthand window onto an important and exciting time in American history.

A few years ago my grandfather was a bit bored and looking for a project. He decided to write up explanations around the letters he wrote home during the war and to publish the original letters and photos, with his annotations, as a book. It soon became clear that the project was too small to appeal to mainstream publishers, and, even if we had found an interested publisher, my grandfather had some particular ideas about how he wanted to put the book together. This was his personal history, and he would not have editors trifle with something so intimate.

We—well, mostly my brother—put the book together in Microsoft Word and self-published it on a site called LuLu.com. LuLu gave the book an ISBN and printed copies on-demand as orders came in. We then listed the book on Amazon.[14] For the last three years, my grandfather's book, *Bulldozing the Way: New Guinea to Japan,* has sold a couple hundred copies a year.[15] At this point, we're not even sure how people find it; presumably they're searching Amazon or Google for something about World War II, soldier's letters, and the Pacific. Yet the experience of authorship has connected my grandfather to a wide range of people interested in his stories—a great gift for him in his nineties. Recently, a stranger e-mailed to ask for his home address. The stranger knew my grandfather had written a book about his time in the U.S. Army's 1897th Aviation Engineering Battalion, and the stranger had found a mug at a garage sale with his division's logo on it. This was no small feat, as the army created the division for World War II and officially disbanded it thereafter; the division only existed for about two years. Who knew it had made mugs? The stranger sent my grandfather the mug, and he proudly drinks coffee out of it every day.

Million Dollar Laughs

Of course, my grandfather has no pretentions to being a professional artist. In diverse areas of culture, individuals who do aspire to make money from their art are embracing and inventing new routes to commercial success. In 2011, Louis CK was a fairly successful comedian with shows on Showtime and HBO. Then he did a live stand-up comedy show and sold it online, himself, directly from his Web site, for $5 a download.[16] The video file had no restrictions; fans could play (and copy) it anywhere, again and again. Louis CK wrote a short opening plea asking people not to copy and share the file without paying for it:

> Please bear in mind that I am not a company or a corporation. I'm just some guy. I paid for the production and posting of this video with my own money. I would like to be able to post more material to the fans in this way, which makes it cheaper for the buyer and more pleasant for me. So, please help me keep this being a good idea. I can't stop you from torrenting; all I can do is politely ask you to pay your five little dollars, enjoy the show, and let other people find it in the same way.[17]

In just a few days, Louis CK made over a million dollars—a response far exceeding his expectations. In a follow-up note, he posted a screenshot of the PayPal account, the stunning number ($1,006,996.17), and explained what he was going to do with the money. About a quarter of it, he said, would go to cover the cost of making the special; a quarter of it would go to his staff; a quarter of it he would pocket; and the final chunk, a total of $280,000, would go to a handful of charitable causes. The move to "go direct" proved so successful for Louis CK that another comedian, Aziz Ansari, followed suit, using his fame from his role on NBC's *Parks and Recreation* to sell a stand-up comedy special directly from his Web site. In an interview with GQ magazine, Ansari explained, "If you're lucky to have a huge fan base already, you can definitely do this."[18]

It's easy to see how artists and performers who have already built large audiences through traditional media can subsequently use the power of radical connectivity to bypass big companies and gain greater control over their work, income, and careers. Radiohead spent years honing their craft and building a fan base while nurtured by a big recording company. Cash advances from their label gave them time to experiment and create albums, while access to top musical talent enabled them to polish their sound, and marketing and radio play enabled them to gain exposure before a large audience. Will musicians and other creative artists manage to pursue successful careers when "Big Fun" goes away? That's not clear. As Jason Gots writes, "In order to survive, many of today's performing artists and musicians need to become an industry of one, creating and managing diverse revenue streams from live appearances, merchandise, rights, and royalties. This demands a rare mix of artistic talent and business acumen (see Ani DiFranco— singer/songwriter and founder of Righteous Babe Records, for a precociously pre-digital model), or the help of some really good friends."[19]

The Great Unknowns

For a glimpse at the considerable uncertainty now surrounding artistic careers, consider the experience of Andy Eggers. When I first met him at Harvard, he was finishing his Ph.D. in government. We worked together on a project he founded to make mutual fund proxy voting more accessible (http://proxydemocracy.org) and became good friends. One day, after I had known Andy for several years, I was surprised to encounter on Facebook an appeal to contribute to a band called The Great Unknowns, of which Andy was a member. As I watched a video of the band, I was stunned. The band had started as a hobby but had grown into something much larger. They were good— really good—and played original music. Soon their first record was picked up by the small independent label Daemon Records and they toured with the Indigo Girls.

Not everyone in the band was ready to play music full time; Andy had become a professor at the London School of Economics. In a June 2012 interview, he said, "our first album didn't sell enough to attract any labels, but we wanted to make another album." So the band went to a site called Kickstarter, which enables artists to crowd source funding for their ventures. The band raised $8,612 from 149 backers, a sum that would let the group get together and professionally record and release an album. It's a perfect story about the End of Big: a small band with a small audience that creates music without any of the normal channels for production, distribution, marketing or monetization. "Kickstarter helped us have it both ways, in a sense: we could do a serious album, somewhat harboring the hope that it would arouse a lot of interest and lead to great new opportunities, but keep up other jobs. As it turned out, it didn't really change any of our lives (so far at least)." Self-financing and hitting up friends is probably what bands have always done when they want to record something but don't have a record deal; what has changed is that Kickstarter (like other Web resources) reduces the search and transaction costs for hitting up friends (and in some cases others who find out about it), including the cost of distributing the rewards that incentivize the donors or investors.

And it's hardly a unique story. Among the over 26,000 projects successfully funded on Kickstarter since its inception in 2008, a third have been music albums, another third film or video, and about a tenth writing and publishing projects.[20] Not one of these projects needed a big studio, big record label, or big publisher to back them.[21] Approximately one-tenth of the films premiering at Sundance Film Festival in 2012 were at least partially funded on Kickstarter, leading David Carr to remark in the *New York Times* that "at Sundance Kickstarter resembled a movie studio, but without the egos."[22] And Kickstarter is just one of several crowd-sourced funding sites. Others include Indiegogo.com, focused on funding indie films, and PledgeMusic.com, for musicians.

After the Great Unknowns' second album, Andy went back to academia. Kickstarter had given them enough funding to make another album but not enough to cover the marketing and touring expenses the band would accrue when promoting the album professionally. All the band members have day jobs but play music seriously. Without the support of big labels, they and others of their generation are building small audiences and scraping together the money to pursue their careers. The nature of commercial music making has changed, clearly, and with it traditional artistic career paths. Talented individuals are inventing and imagining what kinds of career paths are possible. For some people and projects, the new conditions of artistic production are just right; for others, not so much.

The Fate of the Big Show

While we can envision individuals like Louis CK and small musical acts like The Great Unknowns making it financially without big companies behind them, not all entertainment we know and love is poised to survive the End of Big. Many movies and television shows are necessarily big—they take a lot of money to make, and not just to pay stars' salaries and giant executive payrolls (although there is plenty of that, too). A single episode of the hit AMC cable show *Mad Men* costs approximately $2.3 million to produce.[23] That cost makes good business sense for AMC, thanks to advertising and cable fees.[24] To date, we simply haven't seen an online video with the production quality, appeal, and intensity of popular television hits or blockbuster movies. Popular online shows like the Shaytards do enjoy large followings, but they lack the gripping appeal of a period drama like *Mad Men*—a production arguably too expensive for the economic dynamics of YouTube.

The one entertainment genre that remains stubbornly in the age of big media is sports. No substitute exists for watching a live major league sports game on television or, even better, in person. The giant reach of professional sports shows no signs of slowing down or submitting to the audience fragmentation and revenue shrinkage other entertainment has seen. The Super Bowl remains the single most efficient way to reach almost every single household in America at once.[25] But a funny thing is happening as professional sports leagues try to take advantage of the digital age.

In 1960, the American Football League (AFL) signed a five-year television contract with ABC that shared the total revenue from the deal across all the teams equally. This revenue-sharing approach has stayed more-or-less intact; it helps teams in small media markets (like the Green Bay Packers) retain payrolls that compete with teams in large media markets (like the New York Giants). Major League Baseball eschewed the revenue-sharing approach, and as a result teams in large media markets (like the New York Yankees) have gigantic payrolls, while teams in smaller media markets (like the Milwaukee Brewers) fall behind. Observers generally see this disparity as bad for the sport; the revenue sharing in the NFL keeps some parity in talent across all teams, while the lopsided economic equation in MLB enables a handful of teams to always buy the best talent.

When the Internet came along, MLB was determined to do it differently and take a revenue-sharing approach. It created a single entity, Major League Baseball Advanced Media (BAM), to manage the league's and each team's online presence. The league hopes that BAM's profitability will lead over time to a revenue share that will help adjust the dramatic existing disparities across team payrolls. Interestingly, MLB's online approach allows for a wide range of ways of watching or listening to a live baseball game online or through mobile applications. While most other sports have had limited success online, MLB's digital strategy has proved relatively lucrative, grossing an estimated $620 million per year (although this still falls short of baseball's overall television value, estimated at $2 billion per year).[26]

While it's hard to see either high-quality television shows like *Mad Men* or live sports becoming a significant source of digital revenue in the near future, some experts think that day is coming. In May 2012, I interviewed Josh Cohen at *Tubefilter*, an online publication covering the online video world. As he told me, "Online viewership is ahead of where the advertising dollars are. We can expect to see the quality of shows—including the star power—increase as advertisers catch up with the audiences." Cohen points to the appearance of branded entertainment reminiscent of the early days of radio dramas—brands like BMW financing high-quality online videos, both ongoing shows and one-time short films. In my view, it's hard to imagine wealthy patronage in the style of the Renaissance leading to *Mad Men*. Although movie and television studios are experiencing dramatically falling revenue on all fronts,[27] these revenues remain at a scale dramatically greater than anything online video currently offers or is likely to offer. The online audience is fragmented, spread across a much wider volume of online video, and distracted by other activity, including creating and uploading online video. And this range means that for the foreseeable future, we might have to do without high-quality shows like *Mad Men* if traditional Big Fun companies continue to lose revenue.

Not that high quality is totally absent from the Internet. NBC launched Hulu.com to try to capture an audience seeking high-quality online video, primarily by using content from its broadcast and cable offerings. Netflix has licensed a wide variety of television and movie fare, but it has also recently undertaken a multiyear, multi-billion-dollar bet on original programming, wooing major talent away from HBO and Showtime. It seems that YouTube plans to ignore Hollywood and focus instead on cultivating native talent to remain competitive. The company has announced a $100 million fund to invest in people and companies creating content exclusively for YouTube. As YouTube has been forced to crack down on piracy, and as competitors like Hulu and Netflix have offered substantial, high-quality television and movie programming, it seems clear that YouTube has felt some pressure to increase the volume of its high-quality programming.

Is YouTube's original content comparable to Hollywood fare? Ze Frank was an out-of-work neuroscience grad living in Brooklyn when he decided to make a short Web video, five days a week, for a full year. The hilarious, oddly compelling online show, *The Show with Ze Frank*, gained a rapid cult following, in large part because of the way it engaged the audience in participatory activity. Ze hosted strange competitions like "Dress Up Your Vacuum," where he encouraged people to dress their vacuum cleaners like people and send in photos, or "Power Move," where he asked people to submit short videos of their "power move." He went on to play chess against his audience, allowing his audience to collaboratively choose each move against him.

Ze launched his show in 2006, before YouTube was widely available and well before it had any kind of revenue-sharing component. Ze struggled to make money with the show and eventually gave it up, moving on to other endeavors. In early 2011, Ze decided to return with *A Show with Ze Frank,* launching his effort with a Kickstarter project that had a $50,000 threshold. It quickly topped $150,000, and Ze was soon releasing videos again regularly—although this time on YouTube, with the benefit of YouTube's Partner program for revenue sharing on top of his Kickstarter fund.

It's hard to imagine Ze's show working on television; it is a quirky combination of video commentary, online community, and scavenger hunt–like challenges. In a spring 2011 episode, Ze read a comment from a viewer asking him to make her a song because she felt sad and alone. Ze provided words and a melody and then asked his viewers to upload audio of themselves singing the song he has written. Over a thousand people uploaded recordings of themselves singing the song; Ze went on to combine the uploaded audio recordings into a single track, which sounded like it was being sung by a crowd: you're not alone anymore.

Watching *A Show with Ze Frank* is not passive; it can even be a lot of work. It's television meets "cognitive surplus"—and it is a model for entertainment that encourages niche community. As we saw briefly in chapter 3, Kevin Kelly has argued that artists—be they musicians, novelists, or filmmakers—need only 1,000 true fans willing to subsidize their work if they are to survive. But the "1,000 true fans" theory doesn't entirely play out in practice; some pursuits are big, risky, game-changing, and expensive. As of now, the online world (and YouTube in particular) seems far better equipped to support quirky, niche acts rather than the big-vision, big-impact entertainment that Hollywood at its best has put out.

So Should We Get Upset over Hollywood's Demise?

Maybe. It's complicated. When I first started writing this chapter, I went looking for people who would speak in defense of the blockbuster. They were surprisingly hard to find. Among industry pros and insiders, a widespread feeling exists that big entertainment has done a pretty crappy job of entertaining us in recent years. Roger Ebert wrote about this, pointing to the unsung popularity of indie or art movies:

> The myth that small-town moviegoers don't like "art movies" is undercut by Netflix's viewing results; the third most popular movie on Dec. 28 on Netflix was "Certified Copy," by the Iranian director Abbas Kiarostami. You've heard of him? [The industry] can't depend forever on blockbusters to bail it out.[28]

Mad Men notwithstanding, Hollywood seems to have a problem putting out eye-catching original fare. In a blog post titled, "Has Hollywood Lost Its Way?," Jason Sondhi of ShortOfTheWeek .com compared the top ten highest grossing films of 1981 to the top ten highest grossing films of 2011. He found that in 1981, seven of the ten films were original; in 2011, none of them were.[29]

Hollywood won't take risks with unproven story lines, and most of us are fed up with it.

On the other hand, some of the artists I talked to were terrified of the prospect of losing their big-industry backing. The "1,000 true fans" model that worked for Josh Marshall of *Talking Points Memo* (chapter 2) offers as much constraint as it does opportunity. It is hard for creative people to specialize when they have to be their own best advocate and promoter, working social media for the maximum return. Can you imagine an introverted creator—let alone a committed recluse like J.D. Salinger—thriving with the end of Big Fun? Without large companies to help fund and promote artists, creators capable of juggling radical connectivity's many and shifting demands will realize more success than those whose temperaments require more solitude, and our cultural space will change as a result. Rich Burlew has a Web-based comic called *The Order of the Stick*. He wanted to use Kickstarter to print hardcover books of his comic strips, something he'd never done before. He ended up with more than 110,000 orders at a total of $1,254,120—far in excess of his original goal of $57,750. *Wired* chronicled Kickstarter projects that have been mixed blessings for the creators and noted, "[Burlew] says he's spending his days wrangling dodgy suppliers and devoting 'more time to financial spreadsheets than anyone in a creative field should ever have to.'"[30]

In particular, I suspect that radical connectivity makes it harder for a serious, committed artist to develop. In the past, artists would focus on getting published by a major publishing house, or getting signed by a major record company, or getting picked up by a major studio. The career path offered what many artists need: direction and structure. As the chaos of digital creativity makes career options cloudy and obscure, fewer creative people are going to seriously pursue art, and those that do will pursue it in small ways—locally or within unusual contexts—making the highest quality art possibly harder to find and less accessible. Artists have always struggled with financing their work. But the age-old caricature of the starving artist waiting tables to make ends meet makes sense because of the prospect that the starving artist might break through. But without Big, there is no breakthrough—just starving artists who may begin to wonder what the point of art is, other than art for art's sake.

It's difficult to find data proving that sustainable commercial success for artists is getting harder, and it is even more difficult to assess the impact of the End of Big on the quality of our entertainment. I spoke with Hilary Rosen, who was CEO of the Recording Industry Association of America from 1998 to 2003. In a June 2012 interview, she argued convincingly that more money can improve the quality of an artistic product: "Would you rather have a full orchestra, or just a digital sample of an orchestra?" Bigger budgets do not necessarily increase quality, she points out, but they do increase the prospect of quality. For the movie *The Amazing Spider-Man*, Sony Films hired Jim Kakalios, a University of Minnesota physics professor, to write a mathematical equation that would appear several times in the film as central to the story's progression.[31] The result was the "Decay Rate Algorithm," an equation relating to cell regeneration and human mortality that is based on real science. Adding that kind of detail throughout a film is hard to do on a small budget—but possible with a large one.

Beyond its ability to fund works with premium production value, Big Media used to cull through the gazillions of artists out there and find the best for us. Now industry A&R has broken down. While it is exciting to publish your own work (be it poetry, music, or your latest philosophical treatise), the sheer volume of material being created is overwhelming, and true talent can get lost. Cultural authority has also disintegrated (a point we'll cover more completely in chapter 7); we don't have critics out there that people listen to and that have sufficiently important platforms to inform us about art and quality. My grandma giving something a "like" on Facebook isn't the same thing.

New forms of curation are developing, relying primarily on social networks. A crop of streaming music start-ups fill up my Facebook feed, declaring what music my friends are listening to right this very second. As our technology gets smarter, social curation will get better at recommending things I should watch, read, listen to, or play. But a range of new questions arises. For one, do I want social network technology looking over my shoulder, noting my every entertainment choice, and sharing it with my friends, even if it is anonymized and recombined? The *Washington Post* has the Facebook app "Social Reader," which tells me what my friends are reading. An older, well-established intellectual heavyweight that I'm friends with on Facebook recently popped up in my feed as having read an article comparing the breast sizes of female participants in the reality TV show *The Bachelor*. I chuckled—it was a surprise, and I'm pretty sure my distinguished "friend" had no idea that the "Social Reader" was reporting back his every click to people like me.

The Filter Bubble

Entertainment consumption within our new social networking life also increasingly threatens to render us more isolated from one another. In the past, every American watched one of three television channels and read a daily newspaper, even if only for sports scores and coupons. A big, shared public sphere existed in which politicians, policy makers, leaders, and public intellectuals could argue and debate—what we commonly call the court of public opinion. By contrast, with the End of Big we inhabit a "filter bubble" in which our digital media sources—primarily Google and Facebook—serve up content based on what they think we want to read.[32] Even your newsfeed on Facebook is algorithmically engineered to give you the material you're most likely to click on, creating a perverse kind of digital narcissism, always serving you up the updates you want most. Nicholas Negroponte called it the "Daily Me," but it was Eli Pariser who coined the term "filter bubble" in his book of the same name.[33]

The danger of the personalization of entertainment becomes clear when we remember entertainment's traditional social functions. We need quality entertainment not just because it's fun but also because it brings us together as a democratic society. This is one of the few areas of life where we share a common bond—and it's rapidly going away. That's a bigger loss than it might seem. We have frequently in our history relied upon high-quality, big productions to help us work through pressing social and political issues. The 1967 film *Guess Who's Coming to Dinner* offered a compelling view of interracial marriage, which at the time was intensely controversial (the year of the film's release, interracial marriage was illegal in seventeen U.S. states). Oliver Stone's 1987 movie *Wall Street* opened up a public discussion about corporate greed, so much so that the lead character, Gordon Gekko, is frequently mentioned in the context of the current financial crisis. Films and books (like Harper Lee's *To Kill a Mockingbird*) and songs (like Creedence Clearwater Revival's "Fortunate Son") that offer a meaningful critique of our culture help us progress and push the ball forward. Radical connectivity allows creative criticism of our culture to "go viral," but such criticism doesn't have the staying power of a big, blockbuster film or novel that changes the way a generation thinks about a subject or a topic.

Some in the tech world are trying to use social media itself to fight the filter bubble phenomenon. Pariser, the former, longtime executive director of MoveOn.org, has launched a new venture called Upworthy that takes compelling cultural content with a political message and injects it into social network platforms like Facebook and YouTube, to maximize its reach and impact. It's almost as if he's combined cute kittens and political activism in a blender, then poured it into social networks and watched it spread with astonishing speed. David Carr described one Upworthy example on his NYTimes blog:

> There was a remarkable piece of video in which an earnest young man from Iowa talked about what it was like to be raised by parents who were both women. It was an articulate defense of families in all forms and received about 700,000 views on YouTube. Angie Aker, who worked at MoveOn, came up with a new headline: "Two lesbians raised a baby and this is what they got." Five days later, it had 17 million views.[34]

Upworthy tears through the content, releasing several memes a day. Still, we're not talking about a full-length movie or a book; we're talking about an image here, a YouTube video there, maybe a clever animation or infographic. This merging of entertainment and politics is well intentioned, but in many ways it depends on the filter bubble and doesn't actually lead to a national conversation about serious issues. At present, we risk becoming trapped in an individual world without shared cultural space; the big community we once shared as a country is fast disappearing.

The Rise of Even Bigger

The presence of YouTube alerts us to another curious—and disturbing—aspect of the death of Big Fun. While small artists and production companies are going direct, the technology is paradoxically creating some things that are Even Bigger—the platforms like YouTube that all of us use to bootstrap our ideas, companies, candidacies, and local governments. While over 100 video-sharing Web sites exist, YouTube dominates, with more than 3 billion views every day of videos uploaded at an astonishing rate of 72 hours of video per minute.[35] That's pretty Big. Or take Apple. For every song sold on iTunes, the company not only makes money but, more importantly, gets customers locked in to their products. If you want to listen to music, it's easiest to download it from iTunes and put it on your iPod. If you've got an iPod with a bunch of music on it, it makes the iPhone more attractive. Soon you're downloading apps to your iPhone, and when the iPad arrives, you realize you can use your favorite apps on the iPad, too. Before you know it, you've spent more money with Apple than you'd ever planned.

Of course, platforms are not as invulnerable as they'd like to believe. Already, iTunes faces threats from streaming libraries like Spotify, where music lives in the cloud and users pay a monthly fee to listen to as much as they want.[36] Also, the platforms are not simply about top-down control; rather, they allow the many to connect to the many. If your YouTube videos garner a large enough audience, YouTube will share advertising revenue with you. To date, more than 20,000 YouTube Partners like ShayCarl receive regular checks from YouTube for posting and sharing their creations, netting as much as hundreds of thousands of dollars a year, possibly even a few million, although data are spotty.[37] Not too bad for a guy like ShayCarl!

The question arises: Which would you rather have, six major movie studios or one YouTube with 800 million aspiring directors? We're not at a place where we have to choose; right now we get both. But the way platforms aggregate power, even as they produce the Rise of Small, presents some significant downsides. As arguably the world's single largest media source, YouTube can exert influence over which videos get featured on the YouTube home page or category subpages. Since it is a company founded by engineers that was bought by a company that was even more engineering-focused (Google), YouTube tends to default toward editorial decisions made by algorithm rather than by people. What shows up on your YouTube home page is principally influenced by what you, and in some cases your friends, last watched on YouTube. Even if YouTube resists putting a thumb on the scales to influence how you navigate its vast sea of video content, it still wields enormous power by virtue of being the single largest online video source by a wide, wide margin. That's not exactly the End of Big, but it is certainly the end of a certain kind of big, what we might call the "old big" of Hollywood.

The perils of Even Bigger technology platforms that enable and encourage the small become most evident in book publishing. Amazon has come to dominate the sale of books by almost every measure (fully 55% of the books sold in the United States are sold through Amazon).[38] With the introduction of the Kindle e-book platform, Amazon appears to hold the fortunes of authors everywhere in its hands. The company has not been afraid to throw its weight around, insisting on specific pricing for books and using its market share in physical books to drive people toward the Kindle. If you buy physical books, you might buy them anywhere; you have no special incentive to buy them at Amazon over a local bookstore. But if you have a Kindle, then you will likely buy all of your books from Amazon, because if you buy an e-book from any other source, you then have to figure out how to get it on to your Kindle.

The more these Even Bigger platforms control your experience, the more power they have. In a well-publicized case, Apple removed an illustrated graphic novel version of James Joyce's *Ulysses* from the iTunes store because of some nudity—"cartoon boobs."[39] In a contract dispute with Macmillan, one of the largest publishers in the United States, Amazon removed every single Macmillan title from their online store—almost 10% of all book titles sold in the U.S.[40] One Friday evening, the books simply disappeared from Amazon, no longer showing up in search results. No announcement, no warning—just gone.[41] The titles returned a few days later, as Amazon and Macmillan got closer to a contract agreement, but the staggering power of the Even Bigger platform to control our cultural fare had been made known.

On the positive side, Amazon is making it easier than ever for authors to bypass publishers and self-publish both physical books and e-books on its site. Thomas Friedman recently described a visit to see Jeff Bezos, the founder and CEO of Amazon:

> Sixteen of the top 100 best sellers on Kindle today were self-published," said Bezos. That means no agent, no publishers, no paper—just an author, who gets most of the royalties, and Amazon and the reader.[42]

The Even Bigger platform creates enormous wealth for the particular platform in question and creates a little wealth for a lot of people. Even Bigger platforms also enable millions—even billions—of people to participate and connect online in a way that was and in some sense still is beyond our imagination. But as we've seen, the overall effect may be dangerously punitive to anyone hoping to make a living from their artistic work. Will a market continue to exist that supports the most serious artists, the people who in many cases have invested years of their lives to create art only to find payouts dwindling and the life of the artist increasingly impractical? The decline of Big Fun requires the creation of new institutions outside the platforms that support a certain kind of work: that which possesses artistic merit and which the new cultural marketplace does not necessarily reward. Companies like Kickstarter are the beginning of these new institutions, but we need more, and we need them soon. We must figure out how to encourage and compensate creative excellence in the End of Big—otherwise we won't have nearly as much fun.

Digital Feudalism

Ultimately, the challenges posed by the end of Big Fun resemble those of the end of Big News, but with a major difference. In almost every instance—Amazon in books, YouTube in video, iTunes in music—new Even Bigger platforms are leading to a growth in the volume of work (albeit of uncertain quality). In journalism, a similar phenomenon doesn't exist. While more bloggers and citizen journalists are working than ever before, the overall volume of investigative or accountability reporting is shrinking, not rising.

In some sense, we're witnessing an ambiguous, endlessly interesting, transitional period in our cultural history driven by the Even Bigger platforms. Anthony DeRosa, a product manager at Reuters, has likened the present age to a "digital feudalism." Amateur entertainment creators are so seduced by sites like Facebook, Twitter, or YouTube that they put their own time, energy, and money into creating engaging content that benefits someone else—the media platforms—more than themselves.[43] Like serfs during the Middle Ages, these creators don't own the "land" on which they reside; rather, it is "owned by someone else, be it Facebook or Twitter or Tumblr." It is easier than ever (from a technological perspective) to create and distribute creative work—but it is harder to make money from it. Even so, some like Shaycarl manage to break through.

As far as I can tell, the end of Big Fun doesn't carry the kind of dire consequences we'll grapple with in other areas. Consumers have more choice than ever before, even if some of the advantages of Big have been lost. We do need to figure out ways of preserving excellence, shared cultural experience, and the care and feeding of talent. Our old institutions in this space simply aren't up to the task; they're not designed for the networked world. Meanwhile, the Even Bigger platforms move in, seeking an aggressive market dominance that creates new issues (like censorship and accessibility) before we've even had a chance to reckon with what's happened. Artists have always struggled to make their art, and they always will. "He turns not back who is bound to a star," said the great artist Leonardo da Vinci, and sure enough, those in pursuit of creative expression will pursue it even in tomorrow's more difficult circumstances.

5

BIG GOVERNMENT

Ambition and rage all faded
from the air, the air subdued to a new sense
of self, something intimate and sure about the way
it whispers subtle truths neighbor to neighbor—[1]

A few years ago, I read a news article about a vital road in Hawaii that had fallen into disrepair. The road, which ran through Polihale State Park, had become riddled with potholes and was generally dangerous. The state government planned to repair it, but it was going to cost $4 million and would take at least two years to budget the money and get the project started. A group of local businesses that relied on the park for tourism revenue grew frustrated with the slow pace, so they took matters into their own hands, putting together volunteers, equipment, and money. Just eight days later, the repairs were complete.

The technologist Tim O'Reilly cited the story in his essay on "Government as Platform,"[2] and it soon became canonical in discussions about Gov 2.0—a way of embodying the potential of communities to work together ad hoc to be smarter and cheaper than giant, lumbering, bureaucratic government. I was relatively suspicious of the story—it seemed too good to be true—until I saw something in my own neighborhood that gave me pause. Near my house is a playground called Robbins Farm Park. It had an unusual slide that took advantage of a giant sloping hill at one end of the playground. This incredible slide was very popular with kids, but it was getting old and becoming a hazard. The local government removed the slide but, because of municipal budget cuts, had no plans to replace it. A group of concerned citizens jumped into action, using free online tools (including PayPal) to raise the $25,000 needed to build a new slide.

While attending a recent conference involving mayors from across the country, I heard similar stories from around the country. Whether it's roads or school improvements (or budget issues), frustrated citizens are mobilizing connective technologies to organize ad hoc projects, sometimes to supplement government activity, but frequently to replace or preempt it. Such grassroots activity is exciting, yet we have reason to feel uneasy. The sheer speed and volume of disconnected online movements and initiatives threaten to overwhelm our elected leaders and turn government into a chaotic, soupy mess. More significantly, citizen initiatives alone seem inadequate to assure the kind of equality of opportunity and large-scale community citizens in Western democracies have come to cherish; the risk is that they might lead to a country of gated communities and citizens with little sense of connection to one another. We can also question whether small-scale initiatives can ever substitute for certain critical functions we expect of big government, such as regulation of business for safety and quality.

I believe self-government is possible in a radically new way, thanks to radical connectivity, but it's up to us to make sure that our new, small institutions are relevant, robust, and effective enough to provide the structure and order we need. The solution isn't to jettison government or render it small enough to be drowned in a bathtub. Rather, we must design and build new processes for government, inviting participation from across the network provided by radical connectivity while giving leaders the right mix of accountability and room to lead. Fundamentally, we must become more engaged as citizens again, reflecting on our expectations of government and the values we hold dear, coming together behind administrative experiments that might yet enable government for and by the people appropriate to daunting twenty-first-century challenges.

Bigger, Badder . . . and More Irrelevant?

Like other big institutions we've surveyed, government hasn't broken merely because of connective technologies; rather, such technologies have only hastened and accentuated a longer-term process of decay. In recent decades, government has grown bigger than ever, yet also hauntingly irrelevant. Today, more than 800 U.S. military bases exist on foreign soil, and 234 golf courses are being maintained for U.S. generals worldwide—all paid for by our taxes.[3] The tiny state of Rhode Island has an annual budget of over $8 billion, while six technology projects from the U.S. Department of Defense are overbudget by the same amount.[4] In 2005, the GOP senator

Judd Gregg, as chair of the Budget Committee, helped shepherd a $2.8 trillion budget through the Senate. "It's hard to understand what a trillion is," Gregg remarked at the time. "I don't know what it is."[5]

But what does all that money get us? Big federal and state governments are increasingly distanced from our everyday lives. Local governments manage most of the services we see day-to-day, like trash collection, schools, public transportation, libraries, snow removal, water and sewer services, police and fire protection. If I told you the federal and state governments were disappearing tomorrow, you might not even know what that means. And if you needed a road fixed, you might do what citizens of Hawaii did—fix it yourself!

I opened with Hawaii, but the low-hanging fruit as far as bloated, inefficient state government goes is California. One example: California has been waiting close to a decade to implement a new electronic case-management system for its courts. Costs have spiraled out of control, from $260 million in 2004 to $1.9 billion by 2010. It's nerd disease all over again. Technology consultants have run amok, building gigantic, useless technology systems without any accountability. Meanwhile, local communities are making hard choices about education, energy, and mundane but critical details like trash collection.

The breakdown of healthy government function has left Americans sadly cynical and unhappy about governance. A 2012 ABC News–*Washington Post* survey found that a startling 84% percent of Americans disapproved of the U.S. Congress's job performance, while 64% "strongly" disapproved—the lowest approval ratings since 1974.[6] For close to two decades, ABC News has also been tracking America's attitudes about government in the ABC News Frustration Index. Over 2011–12, the index carried its highest scores ever, evidence of mounting and even crisis-level dissatisfaction among everyday citizens about their government.[7] Similar survey work by the Pew Research Center suggests that almost 60% of Americans in 2011 were "frustrated" with government generally, while only 29% said "they can trust the government in Washington to do what is right just about always or most of the time." One bright sign: Only 14% of respondents in the Pew Research Center's 2011 poll claimed to be "angry" at government, down from 23% in 2010, suggesting a possible opening to begin forging some new solutions.[8]

The Groundswell

The trend toward grassroots citizen activism aided by technology is itself a huge mark of how dysfunctional Big Government has become. In their groundbreaking 2004 book *The Groundswell*, Charlene Li and Josh Bernoff describe a "social trend where people use technology to get what they want from each other rather than from institutions."[9] Radical connectivity allows people to bypass giant, slow-moving government institutions as part of the groundswell. When the necessary pressure of the groundswell meets the sheer weight of the federal and state governments, all the end of Big Government requires is a pinch of Grover Norquist's conservative invective (to shrink government small enough to drown in the bathtub) and a dash of liberal frustration with giant corporations blocking progress on issues like climate change and financial reform. In essence, people are tending to take charge on the small, local level precisely because Big Government—state and federal alike—is providing no real answers.

Take government transparency. An older generation of activists might have reformed government from within to make
it more open and accessible, lobbying political leaders or running for election themselves. Matt MacDonald, a resident of
Watertown, Massachusetts, took a different approach. With his
children approaching school age, Matt began to get interested
in the school district, which led him to local government. He
found the town's Web site so impossible to navigate that he
created his own Web site for getting town information, including a tool for reviewing the town budget and a way of tracking
topics and people who participated in city council meetings.
Along the way, he wrote blog posts detailing how you could do
the same thing in your town. And he's started to globalize the
tools he's built, so that you can add your own town at http://
www.nearbyfyi.com.

Matt at least demonstrated an interest in government institutions. In many cases, people are bypassing government altogether. Sites such as seeclickfix.com help communities identify
and address infrastructure problems like potholes, broken
traffic lights, bike lanes, and crosswalks—frequently without
involving the government. In Greece, where the national government is flat-out broke, local citizens are using the Internet
to organize a barter system, even issuing a form of currency.
As I will cover in chapter 8, sites like Freecycle.org and
Shareable.net provide informal networks for bartering goods
and services and sharing resources. You can drive your car to
the commuter rail station, and, thirty minutes after you park
it, someone else picks it up to commute to work, sharing the
car by preset agreement.

Money is under assault from more directions than just barter. A host of alternative currencies are blossoming on the Internet, and one in particular—an open-source project called

Bitcoin—appears to be gaining steam. Bitcoin uses peer-to-peer technology to operate with no central authority, allowing anyone to send "money" (the Bitcoin currency) to anyone, anywhere, at any time, and beyond the reach of governments. Bitcoin enlists participants in the community to manage transactions and issue money; the network, rather than a central bank, collectively creates the money. Lest you think Bitcoin is a nerd pipe dream, many companies—even large, publicly traded ones like LaCie—accept Bitcoin as payment.[10] In the opinion of the tech entrepreneur and journalist Jason Calacanis, "Bitcoin is a P2P currency that could topple governments, destabilize economies and create uncontrollable global bazaars for contraband."[11] Recently, Bitcoin has faced significant setbacks, but it is a promising opening salvo in the advent of alternative, postgovernment currency.

It's possible now to build some of the structures parallel to the government with very little start-up cost—like revenue collection, for example. As people find the current system of government slow and frustrating, they'll increasingly turn to the casual opportunities offered by radical connectivity to accomplish many of the same goals, even to the point of using alternative money like Bitcoin. An astonishing range of tools exist that complement and in some ways could replace government if given the opportunity.

The Limits of Small

The groundswell is exciting, but is it an unmitigated good thing? Not really. In some ways, it makes effective governance more difficult. At a meeting of mayors from around the country that I attended, I kept hearing a common refrain: they were drowning in the volume of direct contact with constituents. In a town of 50,000, you only need a relatively small percentage of the people in town to send you an e-mail, a tweet, or a Facebook message to get inundated. This is a symptom of the failure of our government: people don't know how to engage with elected officials, so they naturally turn to the communications technologies they live inside of every day. Yet it also makes for some chaotic and reactive leadership. Can you execute policy according to a coherent vision, and can you remain responsive to the people who elected you, when you've got a seemingly endless stream of communications coming at you?

Some implications of eroding government authority and power are downright scary. An age-old job of government has been keeping the peace—law enforcement—but radical connectivity is challenging even this basic function. The 2011 London riots were the angry work of young people who organized themselves online using cell phones.[12] Philadelphia has seen an epidemic of disenchanted young men who organize online into flash mobs for the purpose of attacking innocent civilians.[13] In San Francisco, flash mobs have crippled the BART subway system, bringing commuters long delays as a group of hackers fights for unfettered access to the BART controlled wireless network, a fight born out of a struggle related to police treatment of Occupy protestors.[14]

When Big Government gets too powerful, we risk authoritarianism and an erosion of individual autonomy. Whittle it away, though, and you get something else—chaos. Should present trends go unchecked, it is easy to imagine a nightmare scenario of social breakdown facilitated by radical connectivity. In China, flash mobs are forming as part of the so-called human-flesh search engine to dispense vigilante justice to those rumored to have committed crimes. A couple of years ago, the *New York Times* reported the case of a Chinese man targeted for vigilante justice after he'd had an affair and his wife had committed suicide.[15] In the next chapter, I discuss the lawless online communities of the Internet—places like the Silk Road, where you can buy serious weapons. As the blog *Gizmodo* explains: "The Bushmaster M4 is a 3-foot rifle capable of firing thirty 5.56×45mm NATO rounds, and used by spec ops forces throughout Afghanistan. It's a serious weapon. But in the Internet's darkest black market, it's all yours. Who needs a background check? Nobody."[16]

Beyond assuring public order in a physical sense, Big Government has added value by bringing accountability to manifold domains of the modern world. We may need a government totally different from the one we have now, but in the absence of self-government, to quote Federalist Paper Number One, we "are forever destined to depend, for [our] political constitutions, on accident and force." Can you imagine pharmaceutical development and testing without government regulation to provide for consumer safety? What about the size, shape, and best practices of highway design and road construction? Or regulation of air traffic control? Even if our present Big Government institutions don't work well, we still need systems to coordinate our energies and ensure that they serve the public good.

Unequal and Fragmented

There are other problems, too, with the groundswell. Excessive, nonresponsive government might be bad, but government itself exists to assure that all Americans have equal opportunity and that the Bill of Rights persists untrammeled. Start with the gigantic, unimaginable size of the U.S. federal and state governments. Along the way, corporations have spent billions of dollars over the last thirty years to make sure public policy is bent toward their purpose and needs.[17] Add in a healthy dose of citizen frustration, combined with a broken political system that stymies real leadership (as we saw in chapter 3). Now drop in the flexibility and utility provided by radical connectivity. What do you have? More and more gated neighborhoods and communities doing their own thing, without care for the poorer communities down the road. Think back to our opening example of Hawaii: If a group of volunteer citizens repair the road (or the playground), can they decide who gets to use it? Our institutions of government are based on the consent of the governed. If people lose faith in government while relying on emerging technology to provide some alternatives, our existing government will lose its legitimacy. The last time that happened in a serious way, it wasn't pretty; we ended up with the Civil War. Looking at the partisanship we discussed in chapter 3, on my darker days I think that's where we're headed.

A country made up of gated communities might seem farfetched—except that communities across the country are already walling themselves off. Radical connectivity also facilitates the outsourcing of local government functions to private companies, another trend that is hastening socioeconomic divisions and impeding communities from working together. To understand how this works, consider Sandy Springs, a wealthy Atlanta suburb that has outsourced most of its city services. The companies providing Sandy Springs with services are located all over the United States, with a Pennsylvania company managing financial affairs for the city while a Colorado-based firm staffs City Hall.[18] The *New York Times* describes it like this:

Applying for a business license? Speak to a woman with Severn Trent, a multinational company based in Coventry, England. Need a word with people who oversee trash collection? That would be the URS Corporation, based in San Francisco . . . Even the city's court, which is in session on this May afternoon, next to the revenue division, is handled by a private company, the Jacobs Engineering Group of Pasadena, Calif.[19]

By outsourcing all city services, the wealthy enclave of Sandy Springs effectively sequesters themselves from the poorer Fulton County that surrounds it. Sandy Springs goes its own way instead of banding together with other, geographically proximate towns to provide services. "We were being exploited," Eva Galambos, the mayor of Sandy Springs told *American City and County*. "Sandy Springs was a cash cow for [Fulton County.]"[20] Maybe, but as outsourcing of government continues, socioeconomic disparities along geographic lines only promise to increase, leading to a fracturing of the larger community.

Right now, we already have all the fracturing we can take—thanks to cultural trends associated with radical connectivity. Our former sense of citizenship, of belonging to a larger commonwealth, has given way to the "filter bubble" we now inhabit, in which our digital media sources serve up content based on what they think we want to read, creating a perverse kind of digital narcissism.

Eli Pariser opens his *book The Filter Bubble* by describing two friends with similar demographic profiles who each google "BP" in the midst of the oil company's disastrous Gulf of Mexico oil spill. One of his friends gets stock quotes and links to the company's annual report; the other gets news articles about the spill and environmental activist alerts. How do we begin to get a national consensus on critical issues if everyone lives in a filter bubble of information that reinforces his or her beliefs? It's easy to imagine two people looking up a presidential candidate online and getting radically different versions of the candidate's bio and positions on issues. Everything might technically be true but is manipulated to appeal to the specific person reading it. These technologies exist and to a large extent are already in play. In a series of investigative articles titled "What They Know," the *Wall Street Journal* detailed the astonishing extent to which we are tracked and profiled online—and how online material gets tailored to us based on our past activity.

Our filter bubbles are so powerful that they cause us to physically sort ourselves into like-minded groups. The journalist Bill Bishop and the academic Robert Cushing write in their landmark book *The Big Sort*:

> We have built a country where everyone can choose the neighbors (and church and news shows) most compatible with his or her lifestyle and beliefs. And we are living with the consequences of this segregation by way of life: pockets of likeminded citizens that have become so ideologically inbred that we don't know, can't understand, and can barely conceive of "those people" who live just a few miles away.[21]

We risk becoming trapped in an individual world without shared cultural space; the Big Community we once shared as a country is fast disappearing, with implications for democracy and social cohesion. As Bishop writes, summarizing a large volume of academic research: "It doesn't seem to matter if you're a frat boy, a French high school student, a petty criminal or a federal appeals court judge. . . . Mixed company moderates; like-minded company polarizes. Heterogeneous communities restrain group excesses; homogeneous communities march toward the extremes." In the context of the political polarization I outlined in chapter 3, the state of our union begins to look tenuous indeed.

Rise of the Nerd God

If attempts so far to reimagine small government using connective technology seem grossly inadequate, can we apply such technology and the thinking behind it to reimagine our big institutions so that they actually work again? It's a mixed picture, and to help you understand why, I need to take you a little deeper into nerd-dom. It won't be pretty, but please bear with me—there is a golden ring at the end!

Tim O'Reilly is a nerd hero among heroes, a nerd god in the great pantheon of living legends. He is the founder of O'Reilly Publishing, the leading publisher of technical manuals. Virtually every nerd in America—and probably the world over—has at least one O'Reilly book on his or her shelf, and many of us (myself included) learned how to code with the help of O'Reilly manuals. In 2005, O'Reilly penned the essay "What Is Web 2.0?" At the time Web 2.0 was becoming a buzzword, but no one was entirely clear what it meant. O'Reilly solved that problem, laying out a complicated but compelling vision of how the Web was evolving and what constituted Web 2.0 thinking.

A core part of O'Reilly's conception of Web 2.0 was the idea of Web as platform. We've talked in earlier chapters about the role of Even Bigger technical platforms to enable the long tail of the small, bringing about the End of Big. Well, O'Reilly wants to apply this notion of Web as platform to Big Government, too.

The idea of a platform goes back to the early days of computer programming. The truth is, computers understand just one thing: electricity or no electricity. Everything you do on your laptop, on your smartphone, on your car's navigation system gets translated down to electricity or no electricity, represented by zeros and ones. But who wants to write in zeros and ones? Imagine if instead of typing E you typed the letter in zeros and ones, or binary code. It would look like this—01000101. That seems silly and inefficient, but in another sense, it is simply amazing. Every thing you do on your computer—every YouTube video you watch, every Skype call you make, every e-mail you send—is broken down into its requisite zeros and ones and then reassembled somewhere else. As Arthur C. Clarke says, "Any sufficiently advanced technology is indistinguishable from magic."[22]

Software was created in part to offer an abstraction of "electricity–no electricity," so that you could use a mouse or type Shift+e to get E. Computer hardware—the physical parts of the computer that you can touch—would use software to provide efficiency to human beings, so that we would be able to communicate with machines in our language rather than the electricity–no electricity of their language. But as software became more abstracted and more advanced, it faced another efficiency problem. Imagine that you have a word-processing program, like Microsoft Word. You start your word-processing program, and it has to load up instructions (also known as code) that translates each key you type on the keyboard into zeros and ones so that the computer will understand your key strokes. And it has to load up instructions that will display zeros and ones as letters on your monitor, and maybe even code that will send those zeros and ones to the printer so that you can print a hard copy. It loads code that lets it talk to the keyboard, the monitor, and the printer. Then you're done with word processing and you want to switch to e-mail, so you launch program like Gmail or Outlook. But now your e-mail program has to load its own code to communicate with the keyboard, the monitor, and the printer (not to mention the mouse and other hardware). If every program on your computer had to manage the keyboard itself, it would not only be inefficient but would probably lead to major confusion.

The solution was to build a software platform that provided certain efficiencies, chief among them providing a common interface to the hardware. The most common platforms are Microsoft Windows, Apple's Mac OS (operating system), and Linux (the open-source option common on Web servers). As a software platform, Microsoft Windows worries about communicating with the keyboard, the monitor, the mouse, and the printer—so that applications like Microsoft Word, Skype, Firefox, and Photoshop don't have to. The platform provides efficiencies between the hardware and the applications, smoothing the user interface (so you don't have to worry about the zeros and ones) and making the applications more efficient and able to specialize on what they do best—like word processing or Web browsing or any one of a number of things. Every digital device you use, from your mobile phone to the photocopier at your office, has a software platform that brings efficiency to hardware management.

When Tim O'Reilly defined Web 2.0 as thinking about the Web as platform, he was trying to draw a few different distinctions at once. One distinction was thinking about personal computer hardware versus the broader "cloud" hardware of the Internet. Where Microsoft builds Windows to run on top of personal computers, Google builds Search to run on top of the Web server computers in the cloud. Rather than build software for personal computers to bring efficiency to individual hardware, build software for Web sites to bring efficiency to the entire Internet.

A second distinction was thinking about the power of individuals versus the power of hardware. The usefulness of a platform like Microsoft Windows or the Mac OS depends in large part on how powerful the computer is on which it runs—how much memory, how fast a processor, essentially how many bursts of electricity (zeros and ones). But the usefulness of Google Search depends substantially on how many people are on the Internet. The core insight of Google's PageRank search algorithm was that it used the links between Web sites to help establish authority. If a lot of Web sites link to WhiteHouse.gov when talking about Barack Obama, then a Google Search on Barack Obama will pull up WhiteHouse.gov. Every time someone goes online and publishes a blog post, a tweet, or an article with links to other Web sites, they are helping make Google Search more efficient. Compare that to needing to buy a new computer to make Microsoft Windows more efficient!

O'Reilly wrote "What Is Web 2.0?" about a year after Facebook was invented but well before it was open to the public.[23] Facebook takes the lessons of Web 2.0 to heart and extends them in ways that might have been hard to imagine in 2005. One of Facebook's main innovations to date is opening up the Facebook world as a platform that allows applications to be built on top of Facebook, bringing the efficiency of platform design to your social network. Consequently, you can use an app like Plaxo, sitting on top of Facebook, to keep your address book up-to-date automatically by mining your social network. Or you can use an app like TripIt to see where your friends and colleagues are traveling—lest you end up in the same city at the same time on business or vacation. Or when you visit the

New York Times, you can see what articles your friends have "liked" or shared. Facebook is trying to be a platform on top of a platform (the Web) on top of a platform (Windows or Mac OS or Linux on your personal computer).

So Again, What Does All This Nerd Stuff Have to Do with Government?

Well, in a 2010 essay titled "Government as Platform," O'Reilly lays out a new way of thinking about government that applies the radical connectivity provided by personal computers, the Internet, and mobile phones. He starts by describing our current model of government as "vending machine government":

> We pay our taxes, we expect services. And when we don't get what we expect, our "participation" is limited to protest— essentially, shaking the vending machine. Collective action has been watered down to collective complaint. . . . In the vending machine model, the full menu of available services is determined beforehand. A small number of vendors have the ability to get their products into the machine, and as a result, the choices are limited, and the prices are high.[24]

Instead, O'Reilly wants us to imagine government as a platform, on which individuals, organizations, and companies can build services and offerings. He encourages us to think of "government as the manager of a marketplace," as "a service provider enabling its user community." As an example, he cites the Federal-Aid Highway Act of 1956. The U.S. government built an interstate highway system, "a key investment in facilities that had a huge economic and social multiplier effect." Businesses large and small, individuals, local governments, and nonprofits all benefited from the "platform" of the interstate highway system.

Essentially, government as platform presumes that government should provide an underlying infrastructure and then let us build on top of that infrastructure in a wide variety of ways. It does not necessarily mean smaller government—but it does mean the end of Big Government, with many smaller units of government ("small pieces loosely joined," to appropriate David Weinberger's description of the Internet). O'Reilly goes on to describe what government as platform might actually look like:

> Being a platform provider means government stripped down to the essentials. A platform provider builds essential infrastructure, creates core applications that demonstrate the power of the platform and inspires outside developers to push the platform even further, and enforces "rules of the road" that ensure that applications work well together.

The platform establishes the minimal rules needed to efficiently allow a wide range of activity. It creates room for groups of volunteer citizens to do start-up ad hoc projects or for small government groups to provide services in a coordinated manner. But it's not the hierarchical approach of traditional government with attendant checks and balances. A question arises: When you bring fundamental questions of equality, safety, and accountability into the equation, is government as platform up to the task? At present, it's hard to say yes. Government as platform isn't a crazy idea—people have already tried to put it into practice. Unfortunately, the results so far have been less impressive than we might like.

Data.gov

In one of his first acts in office, President Obama issued the executive memorandum "Transparency and Open Government." Micah Sifry describes the brief memo:

> Obama's language was dry, but the message is clear: in essence, he is pointing toward a third way between the stale left-right dichotomy of "big government" versus "smaller government." Effective government, Obama is suggesting, may be found by opening the bureaucracy to direct public monitoring, engagement, and where viable, collaboration.

Within ninety days of Obama's memo, the new CIO of the federal government, an innovative entrepreneur named Vivek Kundra, started building Data.gov. The idea behind Data.gov was to make as much government data available to the public as possible, in the most useful way possible. Americans—especially tech-savvy Americans—would find ways to build useful applications on top of the data, creating value for everyone without costing the government any money. It would exemplify Clay Shirky's "cognitive surplus" at work. Rather than spend twenty-two hours a week watching television, some Americans might put some of that time into building useful applications for their fellow citizens using the raw material provided by Data.gov.

Kundra had good reasons to believe Americans would take advantage of any data the government put online—as long as it went online in a useful, machine-readable format. Before being CIO of the U.S. government, he had been the CIO of the District of Columbia. There, he started a contest encouraging citizens to develop apps for the city. First, he made more than 400 data sets available, many of them in real time. Then he gave away $25,000 in prizes to people who created the best applications for citizens to use. "I expected to get maybe 10 entries," Kundra said, "but we got 47 apps in 30 days." The $50,000 spent on the contest saved an estimated $2.6 million compared to what it would have cost to hire contract developers to do the same work."[25] The apps people submitted covered a wide range of ground; ParkIt, for example, tracked available parking, whereas another app allowed users to submit requests to the city government (please fix the pothole on my street!) and then track the progress of that request through various government processes. Kundra imagined the same kind of roll-up-your-sleeves, let's-do-it-ourselves approach working well at the federal level. And it has—to a point.

Take education loans. If you're interested in student aid from the government, you need to fill out the Free Application for Federal Student Aid through the Department of Education. But as part of the application, the Department of Education requires that you include some proof of your income from the IRS, another government agency. Thanks to the data standards implemented by Data.gov, it's now possible from the FAFSA application to ask the IRS to send the Department of Education a copy of your taxes. Imagine that: one government agency talking to another government agency at your request!

Still, three years into Data.gov's tenure, a core challenge to government as platform is emerging. After spending a lot of time, money, and political capital to build Data.gov, the U.S. federal government now publishes almost half a million data feeds covering just about every imaginable topic. A full 172 government agencies and subagencies participate, releasing their data online. This would appear to be a substantial platform that citizens and the private sector can build upon. And in the entire United States, a country rife with innovation and an appetite for experimentation, where *The $100 Startup*, *The Lean Startup*, and *The Startup of You* were all best-selling titles, a country where an estimated 30,000 new companies are started every year,[26] where more than 600 apps are submitted to the Apple iTunes App Store daily[27]—how many apps over three years have been written for Data.gov?

Just 236 apps have been developed for citizens; 1,264 apps have been developed by government agencies either for use by citizens or by other government agencies. Despite success stories like the Student Aid Application, a large volume of apps doesn't exist, and the average American has not even come close to seeing the impact of this open government in his or her everyday life. Data.gov is not helping government regulate markets, or defend the country better, or spur economic growth better. All things considered, it's a pretty small drop in the bucket. Part of the problem is that the federal government doesn't provide many day-to-day services for most citizens, so the opportunities to develop apps for people are limited—which leads us back to the growing size of government, an institution whose purpose seems to be getting increasingly lost.

The challenges of government as platform aren't just at the federal level. Kundra's successor in Washington, D.C., Bryan Sivak, shut down the Apps for Democracy program, saying, "If you look at the applications developed in both of the contests we ran, and actually in many of the contests being run in other states and localities, you get a lot of applications that are designed for smartphones, that are designed for devices that aren't necessarily used by the large populations that might need to interact with these services on a regular basis."[28] In other words, government as platform leads to what Eric Raymond called "scratch your own itch" technical development:

You build apps to solve your problems. But it turns out that the people capable of writing apps don't always have the same problems—or needs, or opportunities—as everybody else. Even when projects like Code for America (a sort of Peace Corps for nerds, sending programmers into city government) build apps with broad appeal, they're not necessarily addressing the highest priorities—and greatest needs—of local government.

One of Code for America's most popular apps is Adopt a Hydrant. It's a clever Web-based way for people to "adopt" fire hydrants, making sure that in the middle of snow storms the hydrants are dug out and available to firefighters. It was developed in Boston but is being used in Chicago, Honolulu, and Buenos Aires—not always for snow removal, of course! But in the midst of perhaps the greatest budget crunch facing local governments since the Great Depression, I have to believe there are more pressing priorities than snow removal on fire hydrants. To be sure, it's an important and critical task that saves lives. But is it really a crowning achievement for Code for America in a city like Boston, facing a $2.5 billion structural "gap" in its annual budget?[29]

Here's the thing: We actually need the technology to build a better system of government. I love the idea of Code for America—send programmers into places that desperately need innovation and efficiency. The problem is that without a sense of policy, government priorities, or government responsibilities—without digging into the challenges of local budgets and the politics embedded therein—it is very, very hard to effect change using technology. We need political imagination before we can have technical innovation that affects our communities.

Getting to the Heart of the Matter

And that's the heart of the matter. Together, we need to engage in a dialogue about taking control of our destinies again and remaking Big Government. We need to figure out collectively what our values are, how they relate to concrete priorities, and how a revamped government could best deliver on those priorities working for the common good instead of special interests. In some sense, we're in a really good place. The opportunities provided by radical connectivity could change the way government works in the United States. Nothing about our government is immutable. We vote on Tuesday because in the days of horse and buggy, you rode your horse into town to vote on Tuesday because Sunday was church day and Wednesday was market day (see WhyTuesday.org for the whole story). There are 435 members of the House of Representatives basically because that's how many chairs fit in the room, despite the fact that the population size of the average House district has doubled over the last fifty years, from about 400,000 citizens in 1960 to about 800,000 citizens today.[30] Our government today is truly in need of an update that acknowledges twenty-first-century realities and capabilities. If we want to continue living in a prosperous, safe, thriving democracy, we need to start talking about the government we need and want and then take steps to make it a reality.

Does this sound like a daunting task? I think it's daunting in the sense of *exciting*, because it brings me back to our country's earliest days. I love reading about the founding of the United States. I find the Federalist Papers especially inspiring reading. These (white, wealthy) men were trying to figure out what self-government meant and how representative government should work. They had no contemporary models and precious few

historical ones. They struggled with fundamental and important questions—ones worth returning to in the era of radical
connectivity. Are we really doing government the best way
possible, all things considered? That's where we are now. And
if the founding fathers could succeed in creating something
great in their day, so can we in ours.

In reality, history has had precious few examples of democracy. We're a rarefied category—one that could use some more
experimentation if it is to survive in the digital age. In his account of the Greco-Persian Wars, the British historian Tom
Holland writes about the birth of democracy in Athens:

> Athens had become a city in which any citizen, no matter how
> poor or uneducated, was guaranteed freedom of public speech;
> in which policy was no longer debated in the closed and gilded
> salons of the aristocracy, but openly, in the Assembly, before
> "carpenter, blacksmith or cobbler, merchant or ship-owner,
> rich or poor, aristocrat or low-born alike"; in which no measure
> could be adopted, no law passed, save by the votes of all the
> Athenian people. It was a great and noble experiment, a state
> in which, for the first time, a citizen could feel himself both
> engaged and in control. Nothing in Athens, or indeed Greece,
> would ever quite be the same again.[31]

It sends chills down my spine to imagine the ancient Greeks
talking and arguing and finally deciding that they were not
going to choose their leaders based on military victory or he-
redity; they were going to choose them together and hold
them accountable. It was a magnificent moment of political
imagination—and I believe that today we stand at a similar
moment in human history.

Some citizens are starting to ask "heart of the matter" questions about Big Government. In some of our largest states, like California and New York, many believe the state government has more or less failed. Locked in partisan gridlock, unable to manage its own finances, California's government has moved toward a perilous combination of bankruptcy (it has begun issuing IOUs in lieu of payment) and irrelevance. A group of citizens is advocating for a constitutional convention that would wholly rewrite the way California governs itself, potentially moving to a parliamentary system.[32] Inspired by the California effort, a diverse group of individuals—including the progressive Harvard Law professor Larry Lessig and members of the Tea Party—have started the movement Call a Convention (http://callaconvention.org/) to encourage people across the country to have their own constitutional conventions, to reimagine government's very foundations. Iceland has gone so far as to totally rewrite its constitution, electing a constitutional committee of twenty-five citizens who regularly post their ongoing work online and solicit participation from, well, everyone in the country via social media outlets like YouTube, Facebook, and Twitter. As this book goes to press, the new constitution will be in the final round of review by the Parliament of Iceland.[33]

One Answer, Courtesy of the Technopoly

As we begin what I hope will be a lengthy, involved, collective process of political reimagination, there are some paths we should remain wary of. Obviously we should steer clear of political processes and structures misaligned with democratic values that we Americans hold dear. There will be no Politburo or Central Committee making decisions by fiat. But we also need to steer clear of solutions that, although alluring on their surface, apply the basic thinking of the techno-nerds who brought us radical connectivity.

In chapter 1, I explained how nerd history carries an implicit critique of institutions and an empowerment of the individual. John Perry Barlow's manifesto ("Governments of the Industrial World, you weary giants of flesh and steel, I come from Cyberspace, the new home of Mind. On behalf of the future, I ask you of the past to leave us alone. You are not welcome among us. You have no sovereignty where we gather."[34]) is a reminder of the radical mind-set that helped create our current moment in technology and of its potential relevance to government.

In 2009, Wired wrote a chilling profile of Craigslist that described the peculiar attitude of Craigslist's corporate culture—unusual for a company estimated to have a net profit into the hundreds of millions:

> The axioms of this worldview are easy to state. "People are good and trustworthy and generally just concerned with getting through the day," Newmark says. . . . All you have to do to serve them well is build a minimal infrastructure allowing them to get together and work things out for themselves. Any additional features are almost certainly superfluous and could even be damaging.[35]

This way of thinking in software design has a long pedigree—back to the "scratch your own itch" of Eric Raymond's The Cathedral and the Bazaar, but more recently in the best-selling The Lean Startup, whose core admonition is to arrive at the minimum viable product as quickly as possible. It's a compelling vision for running a software company or even an online services company. But does it work as an approach to government? Not so much. As Gary Wolf puts it in the *Wired* profile:

> His cause is not helped by the fact that if the Craigslist management style resembles any political system, it is not democracy but rather a low-key popular dictatorship. . . . Its inner workings are obscure, it publishes no account of its income or expenses, it has no obligation to respond to criticism, and all authority rests in the hands of a single man.[36]

I don't mean to single out Newmark. I've met him, even spent some time with him, and he is an honest, genuine man working hard to do good. Yet I'm concerned with the ideological, anti-institutional thread running through connective technology, from Ted Nelson's "Computer Lib" in the 1970s, to Steve Jobs's literal and metaphorical 1984, to John Perry Barlow's declaration of independence in the 1990s, up to Craigslist today. What if government adopted Craigslist's core philosophies? True to the Craigslist ethos, customer service would be paramount. Newmark famously talks about himself as nothing more than a customer service representative. But this hands-off ethos got Craigslist into trouble as it became the primary online marketplace for prostitution and paid sex—a part of the explosion of the trafficking of women and children across borders that has skyrocketed over the last decade.[37] It's one of the downsides of the hands-off ideology that Craigslist adheres to. On September 15, 2010, Craigslist announced that it was closing all of the erotic ad sections of its Web site.

I promised my wife I wouldn't include any references to science fiction in this book, but I can't help myself. A vision of the future of government structured along the lines of our technopoly's reigning ideology is the conceit behind a clever story by Paul Di Filippo titled "Wikiworld." It begins:

> Russ Reynolds, that's me. You probably remember my name from when I ran the country for three days. Wasn't that a wild time? I'm sorry I started a trade war with several countries around the globe. I bet you're all grateful things didn't ramp up to the shooting stage. I know I am. And the [country] came out ahead in the end, right? No harm, no foul. Thanks for being so understanding and forgiving. I assure you that my motives throughout the whole affair, although somewhat selfish, were not ignoble.[38]

Di Filippo imagines if the United States were run by the same process and approach as Wikipedia, leading to a chaotic but generally fun-loving nation. In Di Filippo's vision, everyone is equally empowered to participate in the system and quickly rises or falls based on his or her ability and interest in investing attention in the system. Consequently, anyone can end up running the country, however briefly, and make decisions with far-reaching consequences just to satisfy some need or desire in their own backyard. It is chaotic but not without its appeal—refreshing in the extent to which it reimagines how our process of government might function. O'Reilly's notion of government as platform is a generation evolved from the "people are basically good, so let them do their thing" assumptions of Craigslist, but it still carries a heavy assumption: that people—individual citizens—will build on top of the infrastructure provided by government to create value for all citizens and solve systemic challenges. Yet the foundation of self-government—at least as imagined by the founding fathers—was about holding power accountable through process. Government as platform may be able to address some of the current power structure's short-comings, and even provide more accountability in the short term, but it simultaneously creates powerful new constituencies that we must also hold accountable in a democracy.

Making Small Work

To do my own part in beginning a discussion about the future of government, I'd like to suggest where I think the real pay dirt is. Although the concept of Big Government as platform is intriguing, the way to better government isn't at the level of big—it's at the level of *small*. The federal government opens up a layer of data—but nothing significant happens, because that's not where the heart of government lives. The heart of government is in our local communities, and that's where we need to reimagine government and rethink democracy and how it should function. That's also where we must begin to understand and imagine new processes for recognizing and protecting the Bill of Rights.

One of the things Alexis de Tocqueville loved best about the United States was that when Americans faced a problem, they would call a local meeting and figure out a solution. He noted that local politics was a central part of the life of most Americans: "The cares of political life engross a most prominent place in the occupation of a citizen in the United States, and almost the only pleasure of which an American has any idea is to take a part in the Government, and to discuss the part he has taken."[39] This is not the American of today, where many local elections across the country have single-digit voter participation.[40] By empowering individuals, radical connectivity opens up new possibilities for local government at a time when our politics—and our government—have become hopelessly focused on Washington, D.C. That's where the end of Big Government can offer an opportunity to reimagine our democracy, taking seriously Jefferson's admonition that we rewrite the Constitution every generation.

Jefferson wrote several times about the need for each successive generation to roam free of the laws of the previous generation. He believed that each generation had "a right to choose for itself the form of government it believes most promotive of its own happiness."[41] Power becomes ossified over time; it takes on weight. Self-government is first and foremost about process. Americans don't vote because they don't feel any ownership of the current process; reimagining local self-government, rewriting the Constitution given the realities of radical connectivity, could reawaken the long dormant politics of our local communities and give Americans a reason to participate in government again. Indeed, the kinds of technology-based local initiatives I described at the beginning of this chapter suggest a kind of instinctive grasping for real solutions precisely where these solutions will ultimately emerge—in geographic settings and on a scale where we can see, touch, and feel the solutions.

The signers of the Declaration of Independence were giddy at the notion of freedom from a hereditary ruling class and buoyed by the democratic town meetings of agrarian New England. But despite their faith in the "meeting" culture Tocqueville describes, they understood the tendency of power to accrete and the corresponding need to hold that power accountable. They designed the Constitution to try to guard against the centralization of power, to encourage a grassroots political culture, but also to protect against the chaotic, narcissism-fueled leadership that can emerge from populism (as imagined in Di Filippo's story). *"The Founders feared both the Monarch and the Mob,"* writes Howard Fineman. *"Now the salving, balancing middle is being ground to dust between the two"* by a combination of global capital and the Internet, resulting in *"a country paralyzed by social and economic as well as political division."* Unfortunately, we're not yet putting much imagination into crafting alternatives at the level of big government. We need to imagine—and bring into being—new processes, new institutions, that on the one hand recognize the core values of our country's founding but by the same token take advantage of the radical connectivity that is a reality of the digital age.[42]

Becoming Citizens Again

In late October 2011, a Republican congressman named Lamar Smith introduced the Stop Online Piracy Act, or SOPA for short. The bill slowly accrued support and looked likely to pass. It was, for the most part, written by entertainment industry lobbyists concerned about digital piracy and its impact on the entertainment industry, especially in the realm of television and movies. But some Internet companies were concerned about provisions in the bill that seemed onerous and would potentially kill free speech online—like a provision that seemed to say that anyone could file a complaint that would require a Web site be shut down, so that a single pirated video on YouTube might lead to YouTube being taken off-line with little warning. In mid-January, an online day of action led by the online community Reddit (owned by Condé Nast) snowballed. Wikipedia joined the day of action; so did Google. Visitors to those sites were invited to send a message to Congress: Stop SOPA. Google alone collected more than 7 million signatures in a petition to Congress; Wikipedia estimated 160 million people—virtually every American who went online during that time period—saw the call to action.

Such mobilization stunned the political establishment in Congress. Here was a sleepy bill, working its way through the process, when suddenly it seemed like the entire Internet woke up and went on the attack. Needless to say, Congress rejected the bill quite definitively. Almost one hundred members of Congress who had previously favored the bill changed their minds overnight. At the bill's heart lay some interesting and fundamental issues about how we think about the Internet and creators of media. But one thing in particular caught my attention. The phrase "We are all lobbyists now" came into broad use. I had heard it before; the founders of TweetCongress .org began using it in 2009 to describe their work. But in the context of legislation being created, the idea seemed sick and disturbing: Is this what our government has come to? Citizen as lobbyist? It is a phrase that assumes a kind of corruption of the process, an inability of the average citizen to get through to their elected officials unless elevated to the level of lobbyist. At the same time, it demonstrates a perverse lack of understanding about how the process of self-government should really work.

Rather than talk about ourselves as lobbyists, let's talk about ourselves as citizens engaged in self-government. Doing so means, first and foremost, experimenting. In chapter 3, I mentioned the Pirate Party, with its growing presence in Western Europe. One of the more interesting experiments of the Pirate Party is LiquidFeedback, a project that attempts to build software that facilitates direct democracy and decision making as an alternative to representative government. Although it is not widely used and remains one small software project,

"Liquid Feedback" represents the kind of experiments in networked democracy that we should be doing more of. Let's get back to basics and figure out how we're going to manage a giant, more or less unmanageable country—and economy—with the help of our new radical connectivity, keeping in mind the hard-won values of our Bill of Rights, as well as the spirit of the original Athenians in their pursuit of democracy.

6

BIG ARMIES

Try to praise the mutilated world.
Remember June's long days,
and wild strawberries, drops of wine, the dew.[1]

We've seen that radical connectivity is rendering traditional, big government increasingly irrelevant, pushing cumbersome, expensive, and in many respects inefficient bureaucracies closer to the brink. Yet we haven't spoken of perhaps the most fundamental function of government, the nation-state's ability to defend against external and internal threats. Here radical connectivity becomes scary indeed.

I first confronted the security risks posed by radical connectivity one fine, sunny morning in September 2001. It was my third week living in New York. As I sat down at my desk on the thirty-seventh floor of a building in lower Manhattan, I heard a loud noise, and the woman two cubes over from me screamed. We were too close to the World Trade Center to actually see what had happened, but clearly something had. When the second plane hit the South Tower, many of my fellow office mates packed their bags and left. I descended to the ground floor, and finding the streets full of falling debris and terrified crowds, made my way back up to the thirty-seventh floor, where it was quieter and calmer. A short while later, the towers collapsed. As I watched a giant plume of dust and debris pour through open windows near my desk, my building shaking violently, I was pretty sure this was it. When the dust cleared, I was stunned to find myself still alive, lying on the floor. It took me a while, but I managed to get out of the building and back to Queens, where I was living with my aunt and uncle.

Decades from now, historians are sure to see September 11, 2001, as the moment when the basic calculus of our national security shifted. The destructive power available to the wealthiest nation-states—nuclear weapons, missiles, vast quantities of conventional arms, hundreds of thousands or millions of professional soldiers—used to assure the nation-state's continued power. Today, national security is fragile, with power shifting to technologically equipped terrorist groups, revolutionary movements, criminal enterprises, murky collectives such as Anonymous, and even isolated individuals with an Internet connection. We might cheer when Internet-savvy opposition movements overthrow oppressive, authoritarian regimes, but overall radical connectivity sows chaos and instability, undoing the traditional advantages of powerful militaries. With Big Armies (both good guys and bad guys) fighting to a standstill against ragtag but tech-savvy groups, we must take a cold hard look at our military-industrial complex and reconsider some previously unassailable assumptions of military might. Our approach to national security and to the stability of the nation-state needs to fundamentally change if we are to reckon with the realities of the digital age.

Fighting Al Qaeda

Before 9/11, Cold War thinking maintained a decisive influence over America's institutional approach to national security and foreign policy. George W. Bush's national security adviser at the time of the attacks and later his secretary of state, Condoleezza Rice, specialized in the Soviet Union and was steeped in the Cold War's mind-set and intellectual history. She and like-minded colleagues conceived the primary threat to our national security as another nation-state deploying a military funded by tax dollars. September 11 revealed a new enemy: the nonstate actor. Within a matter of months, "nonstate actor" went from an obscure academic term used mainly to describe the role of multinational nonprofits in international relations to a mainstream term of critical importance. And the greatest nonstate actor of them all—the biggest bogeyman for America—was Al Qaeda.

Al Qaeda's attacks both before and since 9/11 might have exacted a toll, but has connective technology really equalized the playing field between big and small military powers? Let's look at the facts. During the ten years following September 11, 2001, the United States spent an estimated $3.3 trillion fighting the war on terror, responding to an attack that cost Al Qaeda a mere $500,000 to execute.[2] That's not a winning equation for the U.S.; it's a prescription for bankruptcy. While one line of conventional wisdom holds that the Cold War arms race eventually bankrupted the Soviet Union, it appears that in the era of radical connectivity, Al Qaeda is trying with the aid of technology to do the same to us. It's easier than ever to distribute the material required to grow a jihadist movement; social networking, mobile phones, and e-mail make coordination across multiple continents easy and difficult to trace; and the growing potential of anonymous, online black market arms transactions allows for the equipping of terrorists without strong links to established national governments.

Al Qaeda's actual use of radical connectivity has spanned a range of activities, most notably the use of social media to disseminate messages, communicate with followers, and recruit new supporters to their cause. The group has even gone so far as to create jihadi "rap" videos with popular appeal (nothing like reaching out to the youth!).[3] An intelligence aide to a U.S. senator, speaking to PBS *Frontline* on the condition of anonymity, noted, "The Internet is the poor man's television network. Buy a $300 video camera and a PC and you're in business. You can communicate in a very powerful medium almost instantaneously, almost undetectable and free."[4]

Anwar al-Awlaki was a leader of Al Qaeda killed in September 2011 by U.S. military drones. He has been dubbed the "bin Laden of the Internet" for his tireless posts to social media, including hundreds of sermons uploaded to YouTube.[5] Al-Awlaki has inspired a number of terrorists, including the American military psychiatrist who shot thirteen people at Fort Hood, Texas.[6] In one of his most popular YouTube videos, "44 Ways of Supporting Jihad," al-Awlaki encourages supporters of Al Qaeda to become more active online to help disseminate information and grow the movement:

Some ways in which the brothers and sisters could be "internet mujahidin" is by contributing in one or more of the following

ways: establishing discussion forums that offer a free, uncen-
sored medium for posting information relating to jihad; establishing e-mail lists to share information with interested brothers
and sisters; posting or e-mailing jihad literature and news; and
establishing Web sites to cover specific areas of jihad, such as
mujahidin news, Muslim prisoners of war, and jihad literature.[7]

As the Center for Strategic and International Studies (CSIS)
noted, "YouTube videos and online chat-rooms now help disseminate [Al Qaeda and Associated Movements'] ideology to
far-flung audiences, thus reducing the importance of in-person
interaction as a driver of radicalization."[8]

Even before the invention of YouTube in 2005, Al Qaeda and
related organizations were distributing their videos online,
with all the advantages of radical connectivity—global reach,
immediate availability, zero distribution cost.[9] Beyond simply
using the Internet for distributing sermons and other material
advocating their cause, terrorists have used the medium efficiently for training and operational functions, raising money,
and advertising online to recruit and train suicide bombers. The
advertisements encouraged those who were interested to get in
touch via e-mail for training and mentoring: "The aim of this
training is to continue with our brothers who are seeking to
carry out operations that make for great killing and slaughtering
of the enemies of Islam."[10] As far back as 2006, the journal Foreign Affairs reported that, "[t]errorist groups . . . regularly distribute videos online explaining how to make rockets, improvised
explosive devices, and even crude chemical weapons."[11]

In 2008, under pressure from Senator Joseph Lieberman,[12]
YouTube changed its terms of service to forbid "things like in-
structional bomb making, ninja assassin training, sniper attacks, videos that train terrorists, or tips on illegal street

racing. Any depictions like these should be educational or documentary and shouldn't be designed to help or encourage others to imitate them."[13] Terrorists using YouTube were given notice: Tone it down or we'll revoke your YouTube account.

It may seem obvious, but radical connectivity offers significant advantages for organizational communication and coordination. While it's not known exactly how long Osama bin Laden was living in his compound in urban Abbottabad, Pakistan, analysts estimate that he lived there for at least five years.[14] All the while, he managed a diverse global network of Al Qaeda "franchises," depending on the accessibility and reach of radical connectivity. Zachary Tumin, a colleague of mine at the Harvard Kennedy School, has reviewed the seventeen letters the Pentagon has released from the Abbottabad compound:

> The letters span a decade and outline the dimensions of a would-be caliphate—a truly global theater of war conceived, plotted, and executed by bin Laden. . . . The letters show that he was hands-on and prickly about all such organization matters, going so far as to require memoranda of understanding with affiliates.

Writing about Anwar al-Awlaki, bin Laden asks to see his résumé before granting him command of Al Qaeda in Yemen: "How excellent would it be if you ask brother Basir to send us the résumé, in detail and lengthy, of brother Anwar al-Awlaki. . . . Also ask brother Anwar al-Awlaki to write his vision in detail in a separate message."[15] Reports from the AP detail that even though bin Laden lived without Internet access, he would write e-mails and put them on thumb drives that he then passed on to couriers. "The courier would head to a far-flung Internet cafe, send the outgoing messages, retrieve the incoming ones, and then return to Abbottabad with the responses."[16] Without digital technology, managing a loosely connected global terrorist enterprise in today's world would have been virtually impossible.

While the United States has made substantial gains against Al Qaeda and managed to slay most of the movement's leaders, the technological conditions that enabled their rise—and their success—remain stronger than ever.[17] Anonymous online black markets like the Silk Road and its weapons-oriented spin-off The Armory provide an Amazon-style online marketplace (complete with reviews of products and vendors) for the purchase and trading of just about any kind of weapon imaginable. In the Gizmodo article "The Secret Online Weapons Store That'll Sell Anyone Anything," journalist Sam Biddle writes, "With nothing more than money and a little online savoir faire, you can buy extremely powerful, deadly weapons—Glocks, Berettas, PPKs, AK-47s, Bushmaster rifles, even a grenade—in secret, shipped anywhere in the world."[18] Biddle posed as a paramilitary group leader looking to "equip a private army and overthrow a 3rd world government" and found many suppliers willing to outfit him and his crew for a fee. Biddle went on to seek comment from the Bureau of Alcohol, Tobacco, and Firearms, encountering an official spokesperson who seemed to have no idea what he was talking about. While some portion of the community might be scam artists, Biddle concludes that "the whole thing is just too complicated to be wholly fraudulent."

War Games with a "Do-Over"

Want more evidence that small power really is able to take on big power in today's world? The United States' naval power—ten aircraft carrier fleets around the world—underpins its role in assuring the safe transport of all global trade, including oil flowing through the Persian Gulf. The problem is that even aircraft carriers were proved to be sitting ducks a long time ago (remember the 2000 attack on the USS Cole?), and it's a matter of time before the battleship model becomes entirely obsolete, undone by low-tech cell phone communication, speedboats, and missiles. Naval power will be projected through much smaller seacraft, such as the fleet of speedboats developed by the Iranian navy to counter the U.S. in the Strait of Hormuz. While it's hard to see Iran defeating the U.S. in a full-scale naval war, the U.S. is already ordering smaller boats in anticipation of the first sunk aircraft carrier.[19]

For the United States military establishment, the threat posed by "asymmetrical warfare" with nonstate actors should be crystal clear—but unfortunately it's not. In 2002, the U.S. military conducted the Millennium Challenge 2002, probably the largest war games ever conducted in history. The United States, named Blue, faced an unnamed adversary in the Middle East called Red (note the Cold War metaphors at work in the colors). Red was commanded by a retired U.S. Marine Corps general with the totally awesome name of Paul Van Riper. While U.S. forces were equipped with state-of-the-art weaponry, General Van Riper managed to use everyday technology to rout the giant U.S. military machine. With nothing more than some mobile phones, a few small boats (mostly fishing boats), and small missiles, Van Riper sunk most of the U.S. fleet in the Persian Gulf (in part through suicide attacks).

At one point in the games, the U.S. military announced that such tactics weren't fair and changed the rules to allow the Blue boats to be "refloated." Van Riper soon resigned from the game in protest, noting that the game "was almost entirely scripted to ensure a [U.S. military] 'win.'" Vice Admiral Marty

Mayer responded to Van Riper's resignation by saying, "Gen. Van Riper apparently feels he was too constrained. I can only say there were certain parts where he was not constrained, and then there were parts where he was in order to facilitate the conduct of the experiment and certain exercise pieces that were being done."[20] In real-life conflicts with terrorist forces, the United States can't get a do-over.

Cyber Warfare

That was in 2002. Over the last decade, the U.S. national security establishment does not seem to have made much progress figuring out how to preserve national security in the era of radical connectivity. P. W. Singer, a noted scholar on international relations and the future of warfare, has remarked on the foreign policy and national security establishment's stubborn tendency to view the digital world through the obsolete lenses of the Cold War. Applying a range of Cold War ideas and tactics, deterrence becomes "cyber deterrence," and ideas like "flexible response" become enshrined in the new doctrine of "equivalence." In May 2011, the Pentagon announced a formal change of policy: Digital actions can rise to the level of an "act of war," necessitating a military response. In other words, if someone launches a cyber attack, that's grounds for the United States to respond with real bullets instead of digital ones.

As Singer points out, the digital age is not all that similar to the Cold War. The Cold War saw a stark competition in ideology between two nation-states and their allies. The nonstate actors that might use the Internet to launch a cyber attack are multitudinous and often promulgate no coherent ideology. In addition, the digital landscape is deeply interconnected and controlled primarily by private corporations who don't necessarily hold the same concerns as governments (see chapter 9). Singer notes that unlike the coalition of governments in physical space during the Cold War, "the Internet isn't a network of governments, but the digital activities of 2 billion users," making it resistant to Cold War tactics. If a group of hackers were to release a crippling virus into the power grid, the U.S. military would need to be able to identify the attackers and locate them to retaliate with physical military action. What if the hackers live in downtown Berlin but are unaffiliated with the German nation-state? Do we retaliate with bullets and bombs?

It's not that unlikely. Cybersecurity expert Justin Krebs drew attention to an administrator log-in to the U.S. Army Communications-Electronics Command (CECOM) selling for about $500. Granted, the administrator log-in in question doesn't provide much access, but if it works as advertised, it could give anyone the ability to deface the Web site.[21] It took me just thirty minutes to find and purchase the information— what could a motivated, better-informed troublemaker manage to do? Such accessibility of provocative information is unprecedented. As Singer writes:

> The barriers to entry for gaining the ultimate weapon in the Cold War, the nuclear bomb, were quite high. Only a few states could join the superpowers' atomic club—and never in numbers that made these second-tier nuclear powers comparable to U.S. and Soviet forces. By comparison, the actors in cyberspace might range from thrill-seeking teenagers to criminal gangs to government-sponsored "patriotic hacker communities" to the more than 100 nation-states that have set up military and intelligence cyberwarfare units.[22]

In 2008, the National Security Agency discovered a piece of code—a computer virus, essentially—inside one of the U.S. government's most secure, classified computer networks. The code was listening in on classified material and reporting back over the Internet to a sort of digital mother ship. It's not clear exactly how the virus got into this classified network, but it is clear that it came in on a USB drive. Perhaps a U.S. soldier picked up a USB memory stick in a parking lot outside a military base, plugged the stick into his work computer, and without realizing it released a virus into the network that compromised national security. The *Washington Post* reported on another likely scenario for the transmission of the cyber intruder:

> An American soldier, official or contractor in Afghanistan—where the largest number of infections occurred—went to an Internet cafe, used a thumb drive in an infected computer and then inserted the drive in a classified machine. "We knew fairly confidently that the mechanism had been somebody going to a kiosk and doing something they shouldn't have as opposed to somebody who had been able to get inside the network," one former official said.[23]

While the perpetrator of the digital listening operation is still not known—was it another country, organized crime, terrorists, renegade hackers?—Seymour Hersh reported in the *New Yorker* that the U.S. military's response was to order rubber cement plastered over USB ports on government-issued computers.[24] This response seems primitive, to put it lightly.

Some reporters have subsequently linked this episode to the formation of the U.S. Cyber Command, which has itself been criticized for lacking a clear and coherent vision for national security in the digital age.[25] Even more terrifying, the national security establishment has responded in part by trying to assert more control over the everyday online activity of all American citizens. If every fifteen-year-old with a laptop and an Internet connection is a threat, then we need to treat them like threats, or so goes the thinking. General Keith Alexander, the head of the U.S. Cyber Command and the director of the National Security Agency, is quoted by Seymour Hersh complaining about how he is constrained by U.S. government law from spying on citizens: "[General Alexander] has done little to reassure critics about the N.S.A.'s growing role. In the public portion of his confirmation hearing, in April, before the Senate Armed Services Committee, he complained of a 'mismatch between our technical capabilities to conduct operations and the governing laws and policies.'"

Our think tanks and national security policy-making bodies remain mired in approaches that fail to adequately address or even understand the threats posed by individuals with an Internet connection, a laptop, and a reasonable degree of technical expertise. The former Clinton administration antiterrorism adviser and Bush administration cybersecurity czar Richard Clarke sees cyber war as a major unaddressed policy issue; he is primarily concerned with the potential for a cyber war with China. Despite his status in the national security establishment, his book *Cyber War: The Next Threat to National Security and What to Do About It* aroused nothing but ridicule in technology circles. *Wired* magazine's review noted, "So much of Clarke's evidence is either easily debunked with a Google search, or so defies common sense, that you'd think reviewers of the book would dismiss it outright. Instead, they seem content to quote the book liberally and accept his premise that cyber war could flatten the United States, and no one in power cares at all."[26] If we're going to think seriously about the role of radical connectivity in national security, we need to be clear-eyed and resist hyperbole. Yet we also must acknowledge, as Clarke at least attempted to do, that the balance of power has shifted away from traditional militaries toward small groups of sophisticated, dedicated troublemakers.

Recent months have brought the revelation that the United States military, possibly with the Israeli military, has released at least one and perhaps two computer viruses into the world with the intent of crippling Iran's slow march to nuclear capabilities. The first virus was called Stuxnet, and was targeted at specific kinds of machines that would be in use for uranium enrichment. The second virus is called Flame, and it has not been definitively linked to the United States, although the evidence is strong. These proactive acts of "cyber war," while significant programming projects, hardly raise the scale of resource-intensive military operations such as designing, building, and maintaining an aircraft carrier. Comparatively small, nimble teams can carry out cyber war.

But as computer security expert and investigative journalist Chris Soghoian has warned, the way these viruses have been created and released into the wild exhibits a blindness to the unintended consequences of cyber war. Soghoian has suggested that Flame exposes the average citizen to a significant level of security risk on their personal computer by undermining automatic security updates, not to mention the growing consumer privacy concerns.[27]

As the preceding discussion suggests, the term "nonstate actor" has its limitations; it's conceivable that every technically literate person with a laptop and an Internet connection might be able to influence global geopolitics as a nonstate actor. In fact, it's already happening. Consider Bradley Manning and Julian Assange; together they changed diplomacy and, arguably, the governments of several countries—without any exceptional technical knowledge or expertise. Bradley Manning is a computer programmer, but not a technical genius. He was a low-ranking U.S. Army soldier, a private first class, who made use of a fundamental attribute of digital files: They are easily copied and, once copied, easily shared. Manning allegedly had access to the files through the U.S. military's online data-management tools and copied them to share online on WikiLeaks. Julian Assange, a computer programmer and activist, had been working with a team to build WikiLeaks into a known repository for whistleblowers and a trusted source for journalists. He turned the steady supply of material from Manning into wave after wave of disclosures and leaks, starting with a video of U.S. Army soldiers in a helicopter hunting down two Reuters journalists in Iraq. The mother lode was more than half a million "confidential" State Department cables that turned the institutions of diplomacy on their head and caused political uproars in dozens of countries.

Understanding WikiLeaks

WikiLeaks raises intriguing questions about ethics, journalism, privacy, and governance in an age when the Internet has made the mass storage and publication of information practically free to anyone. "Leaks" refers to material that has been kept confidential or secret, with an implied protection of a corrupted power. "Wiki" is originally a Hawaiian word that means "fast" or "quick," but it has come to be used as a noun to describe a piece of software that allows any user to edit a document. A wiki is open by design, inviting collaboration and participation. It is an incredibly powerful tool for information sharing, for use in a wide variety of contexts, from internal wikis to track the history of projects, to wikis used to collect useful information about a product (a manual written by users), to the ubiquitous Wikipedia. Famously, every single page on Wikipedia has a link at the top that says "edit this page." And anyone can click on "edit this page," on any topic on Wikipedia, and make his or her changes available to the world.

Wikipedia has developed a complex and compelling culture, becoming one of a new breed of distributed, network-centric institutions with growing power. It's an odd kind of power, though, because it is distributed and accessible to literally everyone, its only barrier being Internet access. At this point, more than 5% of the world's population—not the world's online population, not the world's literate population—visits Wikipedia on a regular basis (at least monthly). That is stunning reach. But like many of the platforms we've looked at, Wikipedia is composed entirely of small: Everyone in the world who wants to participate in the power of Wikipedia is invited to do so, by clicking that button on every page to make their own edits and additions. Part of the checks and balances of Wikipedia is a radical transparency—every edit, every decision, and indeed every move is published online. These two core values of Wikipedia—that anyone can edit it and that everything is transparent—are part of the attraction of the "wiki" in WikiLeaks. While WikiLeaks actually started as a wiki, it is no longer one. But WikiLeaks is a place that wants to bring radical transparency to the world's largest institutions and prove that anyone—even a lowly private in the U.S. Army—can be powerful beyond imagination.

What is the role of secrecy in democracy and diplomacy, and what does it mean in a digital age where secrecy and privacy have nearly disappeared? It's an important question, because whatever WikiLeaks's future, Internet-fueled leaks are certain to rise again. Assange sees WikiLeaks as a new force in the world, one that works to bring accountability to large institutions like the U.S. government. In a series of blog posts explaining the purpose of WikiLeaks, he writes, "the more secretive or unjust an organization is, the more leaks induce fear and paranoia in its leadership and planning coterie. . . . Since unjust systems, by their nature, induce opponents, and in many places barely have the upper hand, mass leaking leaves them exquisitely vulnerable to those who seek to replace them with more open forms of governance."[28] To Assange WikiLeaks is something between the accountability journalism of newspapers and the transparency activism of the open-source movement. By his count, WikiLeaks has released more classified documents than the rest of the world press combined: "That's not something I say as a way of saying how successful we are—rather, that shows you the parlous state of the rest of the media. How is it that a team of five people has managed to release to the public more suppressed information, at that level, than the rest of the world press combined? It's disgraceful."[29]

Some political leaders in the West have labeled Assange as a terrorist and even gone so far as to call for his death. Former Republican presidential candidate Mike Huckabee called for the head of "whoever" did WikiLeaks,[30] former Republican vice presidential candidate Sarah Palin called for the person behind WikiLeaks to be "hunted down like Bin Laden,"[31] and Vice President Joe Biden explicitly denounced Assange as a "digital terrorist."[32] Others see him as a journalist and a hero. Daniel Ellsberg, who leaked the Pentagon Papers, has been out in front defending Assange: "If I released the Pentagon Papers today, the same rhetoric and the same calls would be made about me . . . I would be called not only a traitor—which I was [called] then, which was false and slanderous—but I would be called a terrorist. . . . Assange and Bradley Manning are no more terrorists than I am."[33]

Micah Sifry, a noted commentator and journalist on technology and politics, regards WikiLeaks as a symptom of a much broader global trend. In his book *WikiLeaks and the Age of Transparency*, he writes, "What is new is our ability to connect, individually and together, with greater ease than at any time in human history. As a result, information is flowing more freely into the public arena, powered by seemingly unstoppable networks of people all over the world cooperating to share vital data and prevent its suppression."[34] At the heart of this trend is the idea—foundational to our nerd oligarchs—that information should be freely available to those who seek to use it, and the open-source approach that such transparency and openness produces not only better software but also better solutions to many problems.

Eric Raymond, in his landmark essay that lays out a new sort of programming philosophy, "The Cathedral and the Bazaar," uses the expression "given enough eyeballs, all bugs are shallow" to describe the power of transparency to bring about accountability. In a computer program, a "bug" is a problem. Closed-source computer programs do not allow anyone except the creators to read their code. Open-source computer programs allow anyone to read their code. If more people are reading a code, they are more likely to bring to light every possible problem or "bug." It's not dissimilar to Chief Justice Brandeis's saying that "sunlight is the best disinfectant." Transparency is an important part of Western democracy; it helps bring about accountability in the institutions of our society. In chapter 3, I outlined the rise of a global transparency movement as a part of the way those outside the political establishment seek to hold political power accountable. Power corrupts, and the history of Western political thought is in large part about the struggle to hold that power accountable in ways that strike acceptable bargains between freedom on the one hand and security and stability on the other.

WikiLeaks certainly inaugurated a new era of radical transparency, one that was a long time in coming for the leaders of our institutions. If you are a leader today, you must assume that every conversation, every utterance, every e-mail or text could find its way on to the public space of the Internet, the "digital commons." The political consequences have been immense and, by historical standards, almost instantaneous. By ripping the lid off institutions, WikiLeaks became something of a catalyst for three very different but not unrelated political movements: the movements collectively referred to as the Arab Spring, the rise of the hacker collective Anonymous, and the broader movement for global transparency that has surfaced from Russia to Chile to Indonesia.[35] I want to be careful not to overstate the role of WikiLeaks and radical connectivity generally in on-the-ground political protest, but they played some role as catalysts and organizing vehicles for a wide range of political protest and related activity. This returns us to our original focus, the physical challenges posed to Big Armies. We've talked about how radical connectivity has empowered terrorists; let's now take a look at how it has enabled civil society to take on established national powers armed with conventional militaries.

The Arab Spring

The Arab Spring evokes the shifting nature of power as well as its ambiguous consequences. For about forty years, more or less since the Six-Day War in 1967, North Africa and the Middle East have had a relatively stable balance of power. A handful of oligarchs, funded in large part by lucrative oil revenues, controlled most of the governments and militaries of North Africa and the Middle East. The United States and Europe helped to negotiate and maintain an uneasy truce between Israel and its Arab neighbors, in part because of the need to continually extract precious oil to fuel the global economy. To stay in power, the oligarchs reigned using bread and circuses mixed with terror, torture, and censorship. Over time the oligarchs aged and, over the last decade or two, found themselves as octogenarians struggling to control populations ballooning with young, frequently unemployed people.

Reform movements across North Africa and the Middle East gained strength, from trade unions to student groups to militant political groups. Pro-democracy and human rights activists were not welcome; neither were terrorist organizations opposed to the U.S. and Europe. The strict control of information inside these countries made it almost impossible to hold power accountable. In this context, WikiLeaks began to publish U.S. diplomatic cables in which U.S. diplomats, including ambassadors, wrote disparaging memos back to Washington, D.C., about the gross corruption of leaders like Ben Ali of Tunisia.

In a particularly memorable cable by the U.S. ambassador to Tunisia, Robert Godec, with memo headings like "The Sky's the Limit" and "Show Me Your Money," the ambassador wrote frankly about the corruption of the Tunisian regime:

> Whether it's cash, services, land, property, or yes, even your yacht, President Ben Ali's family is rumored to covet it and reportedly gets what it wants. . . . Although the petty corruption rankles, it is the excesses of President Ben Ali's family that

inspire outrage among Tunisians. With Tunisians facing rising inflation and high unemployment, the conspicuous displays of wealth and persistent rumors of corruption have added fuel to the fire.[36]

Average Tunisians knew all about the corruption of their leaders. But to have the United States—the world's most powerful country—documenting it with a combination of bemusement, horror, and disgust was a humiliating wake-up call. General dissatisfaction and unrest in Tunisia grew, not to mention political dissension and activism. Mohammed Bouazizi, an unknown twenty-six-year-old in a rural Tunisian town, gave voice to the growing national frustration. After he was harassed into paying another round of bribes to the local authorities, he doused himself with gasoline, and lit a match while shouting, "How do you expect me to make a living?" The nation erupted in protest and rioting.[37] A torrent of political activity followed, much of it facilitated by the Internet and mobile phones. Longtime leader Ben Ali fled to Saudi Arabia, and Tunisia is today struggling to build a new government following democratic elections in October 2011.

How Successful Will the Arab Spring Ultimately Be?

Tunisia was just the beginning. Much as the technology of radical connectivity empowered and inspired the antiestablishment U.S. presidential campaign of Barack Obama and the successful Tea Party challenges to established Senate Republicans, so, too, has the technology assisted a generational challenge to the power establishment in many countries around the Middle East and North Africa. In places such as Yemen, Libya, and Syria, civil war has ensued. Egypt has seen its government overthrown, without the intense violence of civil war, although not without violence. Algeria, Jordan, Morocco, and Oman have seen protests followed by significant governmental reforms. Even Saudi Arabia, in the wake of limited but still unprecedented political protest, has granted women the right to vote (and run for office) in municipal elections starting in 2015.

But just as President Obama and his Tea Party challengers have found radical connectivity a lot less useful in the actual apparatus of the federal government, so, too, have the heady victories of the opposition given way to the complicated realities of governing—not to mention the muddy alliances and uneasy truces that have allowed forty years of uninterrupted development of the oil industries of the Arab world. As I write this, Syria is fighting a civil war; Egypt is hovering in a political no-man's land where the military is preventing the elected leadership from taking control of the government; and Libya is struggling amidst sectarian violence. What will the future hold? Will people eventually take to the streets again? Or will the fragmenting nature of the resistance persist, preventing a coherent governing coalition from forming in each of the nations that have seen their governments fall? The Arab Spring seems to have brought even more instability to an already deeply divided and tenuous region.

It's interesting that most commentary about the Arab Spring seems to take sides, either overplaying the role of technology in helping to bring about the Arab Spring (a "cyber-utopian" view) or overstating how much technology has helped dictators, militaries, and totalitarians terrorize their opponents. Does radical connectivity pose the greatest opportunity for political dissidents or the greatest threat? One thing is certain: It shifts power to the individual, away from the institution. Even Evgeny Morozov, in a book bent on stripping away the fog of cyber utopianism, acknowledges that "new means of communication can alter the size and likelihood of a protest."[38] The end result seems to be a strange new world in which political movements can jump into being but find it almost impossible to form governing coalitions. The ad hoc breakaway group is fully empowered to challenge any establishment—yet can never fully emerge and sustain itself as the establishment.

Mesh Networks; or, Why Dictators (and the Rest of Us) Should Be Running Scared

For the most part, big institutions still control the essential infrastructure of connectivity. At the height of the Arab Spring protests in Egypt, the Egyptian military simply shut down all Internet and mobile phone service for days on end. When service was restored, the authorities "forced mobile operators— including foreign companies like Vodaphone—to send out messages telling people to attend progovernment rallies."[39] I can hear you chuckling now: not exactly the End of Big. Big institutions—mostly governments and militaries but to some extent also multinational private corporations—still control the master switch and can shut down activity, especially activity that threatens stability. Even British Prime Minister David Cameron has suggested that his government should have a master switch. In the midst of violent riots, he addressed an emergency session of Parliament and suggested that social media was to blame for the violence. The police, he said, might need "new powers" to stop social networks from being vehicles of instability and violence. After public outcry, he quickly backed down.

In fact, traditional power isn't quite as strong as it seems, and the implications of the coming chaos provided by radical connectivity are, at best, muddy. On the one hand, the emerging technologies of connectivity and anonymity allow citizens ways to hold even the most egregious governments more accountable. At the same time, fringe groups are able to use these technologies to disrupt and co-opt institutions that might otherwise work towards democracy and stability.

I need to get a little technical on you again. If you have to plug a cable into your computer to get Internet service, there is always the danger that someone will just cut the cable. This happened in Armenia and Georgia in April 2011, when a seventy-five-year-old woman digging with a shovel accidentally severed an important cable. The entire country of Armenia as well as parts of Georgia (where the woman lived) found itself without Internet access for hours. But if you have a wireless connection, it's a lot harder to cut the connection.

Wireless connections do have weaknesses. Mobile phones are wireless, but they have to communicate with fixed phone towers to maintain signal strength. Mobile phone towers underpin power in this day and age; over the last few years, any time a country experiences civil unrest, a fight breaks out over the mobile phone towers. In the aftermath of the Haitian earthquake, the U.S. military stationed guards at mobile phone towers to prevent looters from disrupting these crucial pieces of communications infrastructure.

But another kind of wireless technology does not require towers. It's called mesh networking, and it works something like this: From my mesh network phone, I write a text message to my mother. In a traditional wireless network, my phone would send the message to the nearest tower, which would then route the message to my mother. But in a mesh network, my phone sends the message to any nearby phone that is participating in the mesh. Say I'm sitting next to you on the bus. I write the text message to my mother, and press send. The message jumps to your phone and checks to see if you're my mother's phone. You're not, so it jumps to the next phone. Is this my mother's phone? And on and on—like the classic children's book, *Are You My Mother?*—until the message finds my mother's phone and is delivered. This is actually pretty similar to how Skype works right now in making voice phone calls.

Mesh networking only functions well if a large number of devices are participating in the network. But it's not a remote or new technology. It's been around a while, and it is continuing to improve and change. The One Laptop per Child project out of MIT's Media Lab saw mesh networking early on as a way to provide network connectivity in remote parts of the world lacking traditional network infrastructure. A village in rural Africa might not have mobile phone towers, but you can drop a couple hundred laptops into the local school and immediately have a functional mesh network. Another project, Smesh.org, has outfitted the campus of Johns Hopkins University with an advanced mesh network that works on top of existing devices—you don't need any special mesh equipment; your current laptop or phone connects to the mesh instead of to the Wi-Fi network.

Many cities and towns have begun to use mesh networks to back up their traditional communications infrastructure in case of emergency. Put a mesh node on every single city bus, and then the buses function as the communications infrastructure for the town instead of the mobile phone towers. It's also easy to build the main components. Amy Sun, founder of FabFi, was an MIT postdoc student involved with an MIT project called Fab Lab that sought to build complex technology out of cheap, everyday items. One day in 2008, she got on an airplane, flew to Afghanistan, and set up a Fab Lab in Jalalabad. She ended up building a communications infrastructure without spending much money or even much time. According to one blog, the constituent parts of FabFi networks recall the TV show *MacGyver*—you can take pieces of wood, plastic, or whatever happens to be around and tack a mesh node to it. "Networks of the type created by FabFi operate independently of government control and can be deployed by anyone anywhere where local infrastructure will not permit a conventional network."[40]

Significant technology challenges remain before mesh networking becomes mainstream—but it is coming. Which brings us back to Egypt and the Arab Spring. Visit openmeshproject.org and the first words you'll read are these:

> On January 25, 2011, when the Egyptian government decided to block the entire Internet, OpenMeshProject.org came to life. . . . Our mandate is to provide a two-click mesh installation on any device, anywhere in the world, to connect citizens irrespective of national boundaries. Our purpose is to provide open and free communications to all people at all times.

The idea behind OpenMeshProject is that you should be able to download an app to your phone, enabling you to leave the mobile phone network and those towers that provide a master switch, to join the leaderless mesh network. The only way to shut down a wireless mesh network is to confiscate every single physical device that might function as a node in the mesh—or at least confiscate enough of them to disrupt the network. The next time a government—be it the People's Republic of China or the U.K.—tries to shut down connectivity to subvert legitimate political activity or to prevent criminal activity, it might find all it does is shift the communication to a digital infrastructure it can't shut down, a mesh network built out of low-cost components, including trash. I find this subversion of authority exciting and full of opportunities for previously unrepresented voices to have a seat at the table. The challenge that it presents is the way it equally empowers violent extremists, like members of Al Qaeda.

Other Reasons Big Governments Are No Longer All That

Of course, repressive regimes have a whole arsenal of digital responses at their disposal beyond merely flicking the "off" switch. Evgeny Morozov highlights the "trinity of authoritarianism: propaganda, censorship, surveillance," noting the opportunities provided by the digital world on all three fronts.[41] But even in these areas, technological trends favor the individual over the institutional. Take surveillance. More and more solutions to anonymize online activity and protect online users from surveillance appear every day. For instance, anyone can go to TorProject.org and download a copy of Tor, a computer program designed to provide complete anonymity to anyone using the Web. Originally programmed by—you guessed it—the U.S. military to protect online communications, Tor has become an open-source project with significant backing from a range of sources, including the Electronic Frontier Foundation. Tor is widely used by political dissidents in authoritarian countries to advocate for democracy. The nonprofit Reporters Without Borders recommends all "journalists, sources, bloggers, and dissidents . . . use Tor to ensure their privacy and safety." According to Tor's own statistics, the network averages almost half a million users a day.

Using Tor is not an ironclad guarantee of anonymity and safety from prying eyes, but as far as technology goes, it's pretty tight. Tight enough that a whole criminal underground has emerged, using Tor to create hidden, anonymous Web sites that act as a giant black market. In about twenty minutes, you can download and install Tor on your computer without much effort and no technical expertise—and a few minutes after that be browsing a digital black market. The most infamous such black market is called the Silk Road. Imagine Amazon.com—complete with ratings and comments—selling everything that is illegal: every conceivable drug, files containing millions of stolen credit card numbers, weapons, child pornography, military passwords, and more.

Anonymous, untraceable private networks—the kind that can be built by technology like FabFi or Tor—are called "darknets," literally networks that exist in the shadows of the Internet. They are increasingly easy to build, approaching a trivial level of technical competency. A tool like Tor is a lifesaver for pro-democracy activists risking their lives for fundamental ideals of justice, liberty, and freedom, but it also makes some of the most egregious pedophiles and terrorists harder to catch and hold accountable. Tor has spawned a terrifying dark side, an unregulated free-for-all where anything—literally, anything—goes, without consequence or concern for human safety and social mores. This, too, is part of radical connectivity's abiding and intensifying challenge to the traditional power and authority of the nation-state.

Anonymous

Talk of darknets leads naturally to Anonymous, that loose af-filiation of hackers who undertake digital actions for vigilante justice. In October 2011, Anonymous managed to take down one of the larger pedophile sites from part of Tor's darknet, posting account details for more than 1,589 users from the site's database.[42] Anonymous has also played a role in many of the episodes detailed in this chapter—from launching massive hacker attacks on the enemies of WikiLeaks, to assisting politi-cal dissidents across North Africa and the Middle East in the quest for more open and accountable government, to engaging in other morally questionable activities. More than any other story from our modern era, Anonymous illustrates the crazy, mind-bending paradox of the opportunities and dangers of an era in which the authority conferred by military power is in decline. Anonymous also offers a glimpse into what future in-stitutions might look like—networked versus hierarchical, with strong cultures, customs, and values, but without a due process in which every individual is equally empowered. Un-derstanding Anonymous is an important part of understand-ing where we're heading in the End of Big.

Anonymous grew out of an online community called 4chan, a Web site founded in 2003 by a fifteen-year-old named Chris-topher Poole interested in anime and (not surprisingly for a fifteen-year-old boy) porn. At first, 4chan was a place to post pictures, but it has evolved into a place where anyone can come and talk about or share anything. At any given moment, hundreds of thousands of people will be on 4chan.org at once. It's an online home to millions of people and the birthing ground for many an Internet meme, including Lolcats, "proba-bly the Internet's top meme—the hundreds of thousands of pictures of cats that float around every corner of the Net, with cat-speak captions: 'om nom nom goes the hungry cat.'" Other noteworthy 4chan accomplishments include lodging a swastika on Google's list of breaking trends and spreading a "rumor that Steve Jobs had a heart attack," causing Apple shares to plummet.[43]

In 2008, as 4chan.org was approaching its fifth birthday, it began to develop something of a political consciousness. An internal Scientology video featuring Tom Cruise was leaked, and when the Church of Scientology took legal action against many online communities to suppress the video, the 4chan.org community fought back. What started as an effort to foil Scientology's efforts mushroomed into a much larger effort, including a massive attack on the Scientology Web site. Administrators of 4chan.org started policing their forums more tightly, so a group of 4chan-ers broke off and created a splinter group called Anonymous to continue the crusade against Scientology. Anonymous takes its name from a crucial part of the 4chan.org culture: Everything is temporary and anonymous. There are no archives, no ability to search the site. In contrast to Facebook, which demands that your online identity be synonymous with your real-life identity, 4chan.org encourages anonymity.

Scientology remains an ongoing Anonymous target, with regular protests organized in front of Scientology offices worldwide. Longtime critics of Scientology are stunned by the influx of people and supporters from the Anonymous effort, even as they plead with Anonymous to tone down its more destructive tactics. On NPR, Tory Christman, a former Scientologist who is now a vocal critic said, "It feels like we've been out in this desert, fighting this group one-on-one by ourselves, and all of a sudden this huge army came up with not only tons of people, thousands of people, but better tools."[44] Anonymous has since moved beyond Scientology to undertake other political actions, including operations offering substantial technical support to various Arab Spring movements as well as attacks against enemies of WikiLeaks (Anonymous went so far as to hack into PBS.org in retribution for a *Frontline* documentary about WikiLeaks perceived as unduly negative).[45]

Occasionally Anonymous seems to overstep, as when the group threatened the Zetas, a violent Mexican drug cartel. While the details are hazy, it appears that the Zetas kidnapped a member of Anonymous, perhaps unknowingly. When Anonymous publicly threatened to expose massive amounts of gang data, including personal information, the Zetas threatened to kill ten innocent people for every Zeta exposed by the hacker group. Anonymous seemed to announce they were backing down—but the entire episode is shrouded in confusion.[46] It's curious and worth noting that the threat of long jail sentences and the wrath of most of the world's major law enforcement and military entities have never convinced Anonymous to back down, but a violent, beyond-the-law drug cartel's threats seem to have had the desired effect. When two loosely knit organizations that exist in secrecy outside established norms face each other, the one with guns evidently wins.

Who, or what, is Anonymous? That is very, very hard to say. There is no clear leadership, no hierarchy, no particular home. Things happen with Anonymous via a sort of snowballing: one person starts to do something and then reports back via any one of a number of online forums where like-minded folks might be hanging out, and activities gradually take shape, with support gathering from across the Internet. Tens of thousands of people—perhaps even hundreds of thousands—have participated in Anonymous actions, from online hacker attacks to real-world, in-person protest marches where Anonymous members wear Guy Fawkes masks. When Anonymous claims to have struck, it could be just about anyone, so loose is the label.

But even if Anonymous seems an ill-fitting label for a vague movement of people across the Internet, it still carries cultural weight as an embodiment of the changing nature of security and power. Responding to a series of attacks by LulzSec, a splinter group that later appeared to rejoin Anonymous, the security journalist Patrick Gray wrote this on his blog: "LulzSec is running around pummelling [sic] some of the world's most powerful organisations into the ground . . . for laughs! For lulz! For shits and giggles . . . Surely that tells you what you need to know about computer security: there isn't any." Digital technology and radical connectivity grant enormous powers to individuals, almost completely without restraint. When a handful of people with the requisite technical expertise want to wreak havoc, they can. And they're almost impossible to catch. Despite dozens of arrests over the last four years, law enforcement globally appears unable to shut down Anonymous—because there isn't anything tangible to shut down, just a culture giving rise to an identity.[47]

As we've seen, the real threats facing national security these days are not other states so much as nonstate actors. If you're in the United States, a nonstate actor might be a terrorist organization, like Al Qaeda. But if you're a repressive regime, the threats to power come from the opposition of your people. Technologies like YouTube (designed for commercial use) and Tor (designed to help activists evade repressive regimes) empower both sides of the equation—pro-democracy human rights activists and loose networks of terrorists. The paradigms used to understand the world to date—ideas about nation-states, military might, and warfare—are not so useful anymore. Anonymous encapsulates the challenges of our digital-fueled expansion of individual power while inciting some measure of fear and misunderstanding. It seems to operate with impunity while adhering to its own incomprehensible code.

I would propose, though, that Anonymous is a bit more than a culture. It's the beginning of a new institution, part of the massive reorganizing of power happening in our age of radical connectivity. The phenomenon is figuring itself out, trying different modes of operation—including different modes of participation—and trying to build some kind of coherent philosophy. It may not always look that way, and you may not like the institution it is becoming, but it is without a doubt becoming *something*.

The task of imagining new institutions is hard. How do we do things without precedent? Anonymous is like a volunteer police force targeted to entities that pursue censorship and surveillance—the opposite of anonymity. But it exists without the kind of process, context, or companion institutions that might shape it in ways that would improve it, and our world. Anonymous jettisons centuries of intense debate about the purpose of government and the methods of holding power accountable, and yet it seems better suited to the challenges of our era than many of our militaries and related entities. Our challenge is to take the lessons of Anonymous and its emerging model and recraft it so as to ensure public order and sustain the fundamental values of our democratic institutions.

7

BIG MINDS

One asks for mournful melodies;
Accomplished fingers begin to play.
Their eyes mid many wrinkles, their eyes,
Their ancient, glittering eyes, are gay.[1]

A few years ago, I invited my friend and former colleague Zephyr Teachout to speak to students at Harvard's Kennedy School. She titled her talk "Are Universities Really Necessary?" and made the case that the Internet allowed for a new kind of distributed, engaged learning that would transform education and displace established sources of intellectual and cultural authority. Given the role that elite universities have played in shaping the politics and policy of our country (every member of the current Supreme Court went to either Harvard or Yale), the changing dynamics of higher education would disrupt the nature of power in this country and beyond.

Teachout was right. Authority and expertise are already being displaced in the corridors of academia. Cramming a thousand students into a lecture hall twice a week to receive the distilled knowledge of a tenured professor hardly makes any sense when those lectures can be recorded and distributed digitally. The ability to publish anything at any time to any audience at virtually no cost has led to an explosion of educational and research-based resources online, radically democratizing the creation, consumption, and dissemination of knowledge. And this phenomenon isn't limited to the ivory tower. Authoritative knowledge in diverse professional fields—medicine, law, science—is falling away with the advent of online services that allow for easy circulation of (often crowd-sourced) information. The new accessibility of knowledge has many benefits, but we can also wonder whether the quality of knowledge is eroding. How do we certify knowledge about the world when intellectual authority ceases to exist? What kind of collective mind are we left with when all the established Big Minds are gone? Are we doomed to a chaotic, unknowable world of half-truths, rumor, and innuendo?

The Disadvantages of an Elite Education

Like most of the other established institutions this book has surveyed, higher education has fallen into a decay that radical connectivity has only exacerbated. Any number of critics have castigated the academy for growing overcommercialized, top-heavy, and heedless of its traditional social functions. William

Deresiewicz was an English professor at Yale before leaving academia to write full-time. In "The Disadvantages of an Elite Education," he writes:

> Our best universities have forgotten that the reason they exist is to make minds, not careers. . . . Throughout much of the 20th century, with the growth of the humanistic ideal in American colleges, students might have encountered the big questions in the classrooms of professors possessed of a strong sense of pedagogic mission. Teachers like that still exist in this country, but the increasingly dire exigencies of academic professionalization have made them all but extinct at elite universities.[2]

According to Deresiewicz, academic achievement today is measured almost exclusively in terms of scholarly research and publishing, and in the process the actual educating of students has gotten lost. Because of a lack of negative consequences, says the former Harvard president Derek Bok, "neither faculties nor their deans and presidents feel especially pressed to search continuously for new and better ways of educating their students."[3] Harry Lewis, a professor of computer science and former dean at Harvard, laments the university's declining role in shaping its students as citizens and leaders. "Tenure is given mostly for research, in part for teaching, and not at all for interest or skill in helping students become adults. Few of today's professors enter academia as a mission, a noble calling."[4]

In my first semester at Harvard, I met a number of exceptional students as a fellow at the Institute of Politics. One of them wrote an opinion piece for US News and World Report that chronicled his intense disappointment with his Harvard education. "Faculty here are inaccessible, students are unengaged interpersonally, and two-way education is an anathema," he complained, going so far as to call his time at Harvard "one of the least intellectually stimulating experiences of my life."[5] My own undergraduate experience at the College of William and Mary was exceptional, mostly because professors gave me a lot of time and attention—and they were fun. From them, I learned a method of teaching that, at its core, boasts a dynamic relationship between students and teachers, and it's what I love most about teaching. I'm always trying to figure out how they're thinking, and I love figuring out a way to open up a new idea to them. But what I enjoy most is when students advance my thinking, pushing me to new and challenging ways of understanding the world. Unfortunately, such a two-way process of intellectual growth seems a rarity these days.

Meanwhile, the cost of a college education is skyrocketing. Between 1990 and 2010, the cost of four years at an average public university increased by 116%—and that's after adjusting for inflation. Yet during this same time span, the median household income stayed stagnant.[6] Even two-year-college programs have seen costs more than double, rising by 71%. But are universities delivering increased economic value to students commensurate with all that increased cost? Not at all. As the Washington Post has reported:

> The unemployment rate among those with bachelor's degrees is at an all-time high. In 1970, when the overall unemployment rate was 4.9 percent, unemployment among college graduates was negligible, at 1.2 percent. . . . But this year [2010], with the national rate of unemployment at 9.6 percent, unemployment for college graduates has risen to 4.9 percent—more than half the rate of the general population. The bonus for those with degrees is "less pronounced than it used to be."[7]

Many studies and analyses have established a positive link between college education and long-term earning performance and employment. As the *Christian Science Monitor* observed, that's a dated picture: "[S]tudies are like photographs: They record the past. They say nothing about the clear and present danger that the bachelor's degree is losing value."[8] Caryn Mc-Tighe Musil of the American Association of Colleges and Universities says that bachelor degrees have lost value precisely because they are so commonplace: "A bachelor's is what a high school diploma used to be." The Ohio University professor Richard Vedder, writing in the *Chronicle of Higher Education,* relates that "some 17,000,000 Americans with college degrees are doing jobs that . . . require less than the skill levels associated with a bachelor's degree." When 14% of our parking lot attendants have a bachelor's degree or higher, you know something is up with higher ed.[9]

I could go on. A 2011 Pew study finds the broad American public increasingly wary of the cost of college and the value of higher education.[10] Parents still expect their kids to go to college, but they're sure it's overpriced and not certain their children will get a good education. In a particularly stunning finding, only 19% of college and university presidents think that the American college system is the best in the world.[11] Our college and university presidents don't think our system is exceptional, and many of them think it is getting worse.

One prominent tech entrepreneur, Peter Thiel, finds higher education so much of a dead-end that he has offered to pay promising smart high school kids *not* to attend college. Thiel cofounded the online payment juggernaut PayPal and was an early investor and mentor for two Silicon Valley heavyweights, Facebook and Palantir. Each year, he picks twenty people under twenty years old and gives them $100,000 to stay out of school for two years and focus on building their start-ups. Recipients of the "20 under 20" fellowship receive more than money; they get mentorship and a range of other services.[12] Thiel has compared higher education to a dot-com bubble: "A true bubble is when something is overvalued and intensely believed. . . . Education may be the only thing people still believe in in the United States. To question education is really dangerous. It is the absolute taboo. It's like telling the world there's no Santa Claus."[13]

Here's the thing: There isn't one. As demonstrated by the Pew research, most Americans have a tangible sense of higher education's inadequacy—and so we have yet another big institution of our time that has failed to live up to its purpose.

The University . . . Unbundled

While Big Universities have veered off course, radical connectivity has unleashed whole new approaches to education. In 2002, MIT announced its intention to post all the material from its courses (both undergraduate and graduate) on the Internet, making it free and open to anyone; that bold vision in turn spawned OpenCourseWare. MIT now lists more than 2,000 courses, offering a range of resources for each. Some courses post simply the syllabus and an outline of topics, others include videos of each lecture, complete textbooks, and interactive Web-based exercises. As of this writing, MIT's OpenCourseWare project has had more than 122 million visits from more than 87 million unique visitors.[14] That means that slightly more than 1% of the entire world's population has visited MIT's OpenCourseWare project.

A reading list, a handful of lecture notes, and a midterm do not substitute for the power of a great teacher. Enter Salman Khan. Khan was himself a successful student, attending MIT for an undergraduate degree in math and a graduate degree in computer science before matriculating to Harvard Business School. In 2006, while working as a hedge fund analyst in New York, his younger cousin Nadia was struggling with her middle school math class. Khan began to tutor Nadia, who lived in Texas, using a Web-based notepad tool called Doodle as well as Skype video chat. Soon other relatives were requesting tutoring sessions, and before long Khan was recording the videos of his tutoring and posting them online. An excellent teacher soon finds eager students, and Khan's video tutorials went viral. Khan started a Web site, Khan Academy, and now has more than 2,400 lessons posted online. In a given month, 3.5 million unique visitors view 39 million pages on the site to learn more about math and science.[15]

Inspired in part by Khan Academy, one of the most popular professors at Stanford began teaching an online class about artificial intelligence. Sebastian Thrun was stunned when 160,000 people from around the globe signed up to take the graduate-level class, CS221: Introduction to Artificial Intelligence. Approximately 137,000 people ended up dropping out over the course of the semester, but a stunning 23,000 completed the course. "Having done this, I can't teach at Stanford again," Thrun told Reuters. "At Stanford, priority is your research career. . . . That is counter to teaching 100,000 students, who generate 100,000 emails."[16]

Thrun used the experience to launch Udacity, a for-profit online educational company that offers a couple dozen courses, mostly in computer science and math. Anyone can take the courses free of charge, but to earn certification for course completion, you have to pay a fee and pass a final exam. It's not at all clear that these certifications are worth anything—yet. Participating students in the Udacity online community tutor one another, and students who complete courses can post their résumés online. Udacity uses your work during the course and your résumé to see if they can place you in a job at a high-tech company, earning a placement fee in the process. Your expertise—provided in part by Udacity—becomes a vehicle for company revenue.

While Thrun was starting Udacity, a colleague at Stanford launched a competitor called Coursera. Coursera follows a different model, working with top universities like Stanford and Princeton to put their best classes online. As the cofounder and Stanford professor of computer science Andrew Ng explained to Tom Friedman of the *New York Times,* he normally teaches 400 students but he could teach 100,000 through an online course on machine learning. "To reach that many students before, I would have had to teach my normal Stanford class for 250 years."

Start-ups like Udacity and Coursera are just a few examples of an explosion of quality online learning opportunities. Radical connectivity—the ability to reach anyone, anywhere, with high-quality video, at practically zero cost—is leading to an unbundling of the university education, just as it led to the unbundling of the news.[17] In its former bundled state, popular sections of a newspaper—the sports section, say—attracted subscribers and thus paid for less popular sections, such as book reviews. Now such bundling no longer happens, for each section of a newspaper is available individually online from whatever blogs you care to read; an important part of the newspaper's business model thus breaks down. In the unbundling of the university, all your classes and professors are available individually. Just take organic chemistry—nothing else! And it's not just the classes and professors who are being unbundled—it's the relationship between academic research, teaching, and a well-rounded education. All these pieces are being pulled apart and housed in various places. In some instances, they are lost altogether.

Alan Jacobs, a professor of English at Wheaton College, has noted that knowledge as a whole also risks becoming unbundled from another key element of university education, "credentialing." Universities have long enabled students to "learn stuff that they couldn't learn elsewhere—because the experts weren't elsewhere—and to be certified by those experts as having actually learned said stuff." Now that universities are losing control over the dissemination of knowledge, we might wonder how much longer they will determine authoritatively who among us is recognized as having mastered and assimilated that knowledge, and who isn't.[18]

It's in relation to credentialing, in part, that Thiel's admonition of a bubble in education seems most prescient. Elite, private universities conferring their imprimatur on a select group of degree-bearing students are now only a part of the picture. Over the last decade, technology-driven, for-profit education has been one of the fastest growing industries in the United States;[19] today, approximately 10% of all college students are enrolled at a for-profit institution—more than 2 million students. The University of Phoenix, the largest provider of for-profit schooling, brings in more than $4 billion a year.[20] Yet as the education "industry" has expanded, offering an array of certifications and special or nontraditional degrees, the value of such credentials has declined, and the identity of an "educated person" has become more nebulous. A substantial study comparing for-profit universities and colleges with public and nonprofit education found "there are large, statistically significant benefits from obtaining certificates/degrees from public and not-for-profit but not from for-profit institutions."[21] The *Atlantic* summarized the study: "For-profit graduates have worse job prospects and earn less than their peers who attend nonprofit schools. A new study . . . suggests that many for-profit diplomas are literally worthless in the marketplace. This even holds true when you control for student characteristics like wealth."[22]

A number of things contribute to the market value of an institutionally granted credential for individual students, including the institution's reputation and excellence. Not only can students now find proxies for these functions outside academia, but new ways of building reputations entirely beyond credentialing are also taking hold—things like maintaining a detailed, well-researched, well-informed blog on a particular subject area or making significant contributions to Wikipedia over time, building online reputation. Remember that old *New Yorker* cartoon, "on the Internet nobody knows you're a dog"? From a purely reputational point of view, the Internet may think that a Wikipedia user who goes by the name Hoppyh knows more about Abraham Lincoln than famed historian Doris Kearns Goodwin. After all, Hoppyh has made more than 1,339 contributions to the Wikipedia page on Abraham Lincoln—and Doris Kearns Goodwin does not appear to have made any.[23] In a September 2011 event at the Nieman Foundation for Journalism at Harvard, Goodwin was challenged by a member of the audience to join Wikipedia as an editor; she said she was "intrigued and elated" by the prospect of becoming a Wikipedia editor on Lincoln.[24] Even if Goodwin did join Wikipedia, her status as a scholar on the subject for decades would not accord her any special status; she would have the same standing as Hoppyh, perhaps even less influence given Hoppyh's contributions to date.

Another reason to go to college—beyond credentialing and reputation—is networking. Before radical connectivity, institutions of higher education helped students build and maintain relationships, providing a network of people with a shared identity as "old school chums." The power of that social network has not declined, but it is certainly being challenged by social networks that don't carry the institutional weight of alumni networks. Remember that person you met one time at a party who works at that company where you just applied for a job? It's easy these days to find them online and "friend" them. More broadly, informal groups organizing on social networks like Facebook and LinkedIn can offer an enticing alternative to the chumminess of college dorms.

Reviewing Peer Review

It isn't just the definition or identity of an educated person that has come into question thanks to unbundling—the very constitution of knowledge has become vexing as well. In the past, universities have acted as collectors and curators of knowledge, codifying the means of communicating knowledge down through the generations. But the conditions that gave rise to the university—most notably the importance of physical books, and the need to aggregate books and people in one place to facilitate the transfer of knowledge—have evaporated. This in turn has brought the age-old convention for establishing academic excellence, peer review, under severe attack.

We have long depended on peer review to distinguish fact from fiction and also to establish a pecking order inside academic institutions. The idea behind peer review is that other people in your field can assess your work and verify your research, as well as encourage the pursuit of excellence within a given field. In the discipline of history, for instance, peer review will establish whether a given scholar has performed archival research fairly, documented it well, and come to reasonable conclusions. The granting of tenure to professors is built in large part on peer-reviewed publications, while reputations hinge on the assessments of peer-review panels. Large amounts of money—most of the world's government-funded research—are dispersed based in a large measure on peer review.

Critics now perceive the whole system as out of date. In our age of instantaneous global communication, peer review's slow pace seems to hold back scientific progress. The *New York Times* article "Cracking Open the Scientific Process" summarized the critique as follows:

> The system is hidebound, expensive and elitist, [critics] say. Peer review can take months, journal subscriptions can be prohibitively costly, and a handful of gatekeepers limit the flow of information. It is an ideal system for sharing knowledge, said the quantum physicist Michael Nielsen, only "if you're stuck with 17th-century technology." Dr. Nielsen and other advocates for "open science" say science can accomplish much more, much faster, in an environment of friction-free collaboration over the Internet.[25]

Peer review may provide accountability, but it is in many ways deeply flawed and inadequate in the digital age. Peer-reviewed publication takes on average about two years, and many scientific journals costs thousands of dollars a year for subscriptions. Not only that, but if scientific research fails, it usually does not get written up and published. Who wants to publish an article that says, "we tried this and it didn't work"? "Publication bias" is a well-known challenge in academia. A major review of more than 4,600 peer-reviewed academic papers across a range of disciplines and a range of countries found that over the last twenty years, positive results increased by almost 25%.[26] And yet failure is a crucial part of the scientific process. To better figure out what works, you need to know what doesn't work.

Despite the entrenched function of peer review within the academic establishment, online correctives and alternatives have proliferated. Take the issue of reporting research failures. Jean-Claude Bradley, a scientist at Drexel University, runs a lab looking for chemical compounds useful in the fight against malaria. He started an "open notebook" called UsefulChem to catalog the lab's failures as much as their success.[27] By one account, "his lab started testing as many compounds as they could, recording the results in an open notebook that contained no heroic narrative, just daily results. He then started another open notebook that crowd sources the nearly endless question of which chemicals are soluble in which other chemicals. The result is a mammoth spreadsheet of interactions, most of which are nonevents."[28] Such nonevents are actually crucial information in this context, since the kind of chemical reactions that could form the basis of a new medication take place in an inert, nonreactive medium.[29] By providing radical transparency to his lab, Bradley has created a valuable store of accessible knowledge.

Research that used to be found only in peer-reviewed publications is now finding its way out in other ways. Projects like arXiv and Public Library of Science offer academics access to articles and other research. Another project called Research-

Gate, a social network for academics and scientists, encourages research collaboration and provides a parallel track to the hierarchical process of peer review. Significant scientific research is also increasingly accessible to the broad public. Scistarter.com encourages people to post the results of their own research and invites academics to use it as a place to recruit volunteers to help them crowd source research. Visitors to GalaxyZoo can help scientists better map and understand our galaxy by using photos and data provided by the Hubble Space Telescope: "You sign up, take a brief test to see how well you can identify galaxies (it's fun!), and if you can get a high enough score, off you go! You are sent a galaxy image (a program looks at all the data and decides if an object is a galaxy or not) and asked if it's an elliptical, a merger (a product of two galaxies colliding), or a spiral. You click the appropriate button, and the next galaxy is automatically served to you."[30] In the first few months after its launch, 80,000 volunteers classified more than 10 million images of galaxies.

The crowdsourcing of science uncovers fascinating opportunities. Almost twenty years after the Exxon Valdez oil spill, scientists working on the cleanup grappled with 20,000 gallons of oil that still lay at the bottom of the Prince William Sound. Because of Alaska's cold temperatures, oil thickened and hardened to a cement-like consistency, confounding pumping equipment. The nonprofit Oil Spill Recovery Institute opened the problem up to anyone, offering a $20,000 prize for a viable solution. The institute posted the challenge on InnoCentive, a Web start-up that works with companies and nonprofits to create contests for solving significant scientific and engineering challenges. John Davis, an Illinois chemist with no oil industry background, used a common solution in the construction industry to win the challenge. Davis noticed that in complicated construction projects, cement is vibrated to keep it from hardening, and he devised a similar solution for thick oil at the bottom of the ocean.

In a Harvard Business School research paper titled "The Value of Openness in Scientific Problem Solving," academics reviewed several years of Innocentive data and concluded that the success of the InnoCentive approach owed in part to the diversity of the problem-solvers.[31] It turns out that having a wide range of expertise allows for an intellectual cross-pollination that leads to solutions. Scott Page, a professor of complex systems, political science, and economics at Princeton, has written an exceptional book called *The Difference: How the Power of Diversity Creates Better Groups, Firms, Schools, and Societies*, in which he methodically documents how groups made up of people with a lot of different backgrounds outperform groups of "like-minded experts." Page proves that "diversity yields superior outcomes," going so far as to posit mathematical theorems about predicting a group's accuracy in problem solving as a function of the group's diversity. According to Page, it doesn't matter what kind of problem you are solving; diversity helps you arrive at the most efficient solution, faster. Abraham Lincoln knew this—he made his cabinet a "team of rivals." As Doris Kearns Goodwin memorably noted, "Good leadership requires you to surround yourself with people of diverse perspectives who can disagree with you without fear of retaliation."[32]

The End of Experts?

For traditional academic authorities, the diversity wrought by radical connectivity may prove a rather scary thing. We've entered a brave new world in which anyone, theoretically, can contribute to intellectual debates, regardless of their degrees, publications, conference presentations, or other educational accomplishments. The result is an erosion of professional authority, a trend visible in other professional fields. Need legal advice? You can bypass a certified lawyer and check Web sites like LawPivot.com. Online communities also connect patients suffering from similar medical conditions, allowing them to learn about other people's experiences (PatientsLikeMe.com) and medications (MediGuard.org). On the broadest level, we can discern the end of expertise—or at least vetted, certified expertise—in the success of Web sites like Quora, which invites Internet users to post questions and provide answers. When I sampled three random genres—mechanical engineering, pharmacology, and law—I found that the top questions of the day included the following:

- How much electricity and CO_2 does a building save on an average using district cooling (70 levels)?

- Is there a significant difference in effectiveness between Nexium (esomeprazole) and OTC omeprazole (brand name: Prilosec) when treating GERD?

- If a patent application has an independent claim and 10 dependent claims, and the independent claim gets rejected, can I "migrate" dependent claims to the independent claim until enough steps have been added to make the independent claim novel?

The answers to these questions are manifold, information-rich, and enormously useful, making use of the cumulative expertise of Quora's participants. On the other hand, how can we be sure that the answers Quora gives are complete? What do we lose when we give up certified sources of expertise? Can we really feel certain of anything anymore, as our parents did when they visited their doctors, lawyers, accountants, or religious clergy?

The Internet's deprofessionalization of knowledge can prove disconcerting, if not outright problematic, by rendering practical decision making much more difficult. Research by the Tufts professor Lisa Gualtieri has shown that the persistent googling of medical information by patients complicates the relationship between patients and their doctors and can even cause problems in medical care.[33] Based on information we find online, we question our doctors' diagnosis and try to assess their competence, frequently causing ourselves a lot more stress in the process. Gaultieri describes one online consumer who thought she had cancer yet had to wait for a doctor's appointment. She "relentlessly searched the Internet scaring herself 'to death' by the information she found about severe cases." When this patient did receive a medical examination, her doctor told her that they wouldn't know anything without a biopsy—hardly a salve for her anxiety. Finally her biopsy came back negative, easing her fears, but not before her obsessive searching led into what she termed "'the depths of the Internet.'"

Thanks to the design of our medical system, doctors frequently don't have the ability or incentive to spend much time with their patients. The average primary care doctor in the United States has 2,300 patients and is unlikely to remember your name—or even recognize your face—if he or she runs into you in the parking lot of the grocery store.[34] The quick and easy availability of online medical information helps fill the void left by our inadequate medical system, but frequently leaves individuals scratching their heads, wondering what to do, and worrying themselves sick.

The sheer volume of information now available thanks to radical connectivity also stymies public debate by opening up endless possibilities for verifying and disputing a given fact. In a world where my facts can always be disputed by your facts, and we both have equal weight online, scientific consensus becomes meaningless for many kinds of disagreements. Dan Kahan is a professor at Yale Law School who has written widely on cultural cognition. He asks, "Why do members of the public disagree—sharply and persistently—about facts on which expert scientists largely agree?"[35] Based on his research, Kahan argues that our values shape our beliefs, so much so that our values interfere with reason. Instead of asking if something makes sense based on data, we ask ourselves if something fits into our idealized worldview. This helps to explain why arguing with someone about climate change never changes their mind and frequently leads them to sink into an even stronger position. When you dispute their view of science, you're disputing their entire worldview, their cultural context.

This rise of uncertainty about knowledge and truth is part of our political paralysis—the polarization we discussed in chapter 3. All issues become political footballs, and every side can get their own facts to support their position. As a result, it gets harder for objective, accepted truth to rise above the political fray and undergird a consensus policy. The ever-expanding, vast volume of information available online also intensifies the breakdown of communal identity linked to the "filter bubble" effect I discussed earlier. Through personalization technology, not only do you see things optimized to your specific, individual taste; you can also find an endless amount of data to back up your increasingly sectarian viewpoint.

Am I painting too bleak a picture? David Weinberger isn't bothered by the erosion of academic authority. In his recent book *Too Big to Know,* he argues that the state of knowledge is simply "too big" for any one human being or even group of human beings to comprehend.[36] The solution lies in the distributed knowledge of radical connectivity: To understand something, engage in a conversation with people online. Assemble a powerful network of thinkers from across a range of disciplines, and you'll be able to solve big problems. In an interview with *The Atlantic,* Weinberger succinctly articulates his book's thesis:

> My generation, and the many generations before mine, have thought about knowledge as being the collected set of trusted content, typically expressed in libraries full of books. . . . Yet, for the coming generation, knowing looks less like capturing truths in books than engaging in never-settled networks of discussion and argument. That social activity—collaborative and contentious, often at the same time—is a more accurate reflection of our condition as imperfect social creatures trying to understand a world that is too big and too complex for even the biggest-headed expert.[37]

Instead of finding knowledge in a long, deep plunge—like the linear progression of a book—Weinberger argues that knowledge is better off existing in networks, and in fact will exist there in the future. But there's a downward spiral at play here: The world is getting so complex that expertise is losing its traction, which makes for a messy world in which the old arbiters of authority are questioned and compromised, in turn making the world seem even bigger and more complex!

Is Obama a Muslim?

At its worst, the Internet sometimes seems to devolve into a swirling miasma of contentions and rumors entirely removed from actual, real-world fact. Practically anyone can say anything and achieve a reach and impact (however fleeting) that previously might have presumed some level of legitimacy and endorsement. The sad specter of JFK's press secretary Pierre Salinger brandishing a photo printed from the Internet ostensibly "proving" that a missile shot down TWA flight 800 comes to mind. A quick sample of Snopes.com "top 25" Internet rumors turns up provably false gems including that Starbucks has refused to send product to active duty Marines in Iraq, entering your PIN in reverse at any ATM will summon the police, and the spurious claim that automobile components emit cancer-causing benzene fumes.

Weinberger has summarized Cass Sunstein's impression of such rumors as "information cascades of false and harmful ideas . . . that not only gain velocity from the ease with which they can be forwarded but gain credibility by how frequently they are forwarded."[38] Individuals and groups on the Internet might believe in evolution, but they might also decide that Barack Obama is a Muslim, born outside the United States, and thus an illegitimate president. Indeed, it was a similar kind of disorganized online sentiment that fueled significant and unfortunate challenges for Barack Obama's presidential campaign and persisted long into his administration.

According to the *Los Angeles Times,* false rumors saying that Obama was secretly a Muslim started during his campaign for the United States Senate in 2004 and expanded through viral e-mails by 2006. The *Times* compared these rumors to earlier false rumors about the 2000 presidential candidate John McCain fathering a dark-skinned child out of wedlock.[39] In December 2007, the Hillary Clinton campaign asked a volunteer county coordinator to step down after she forwarded an e-mail message that repeated the false rumor that Obama was a Muslim.[40] In June 2008, the New York City mayor Michael Bloomberg, himself Jewish, spoke out to Jewish voters in Florida against false e-mail rumors that said that Obama was secretly a Muslim and did not support Israel.[41] You may have been among those who received a forwarded e-mail full of false information about Obama's background, which included passages like, "Obama takes great care to conceal the fact that he is a Muslim. He is quick to point out that, 'He was once a Muslim, but that he also attended Catholic school.'"

As the *Washington Post* notes, these kinds of e-mails are nothing new: "Nonpartisan debunkers such as FactCheck.org, Snopes.com, PolitiFact.com, Emery and the *Washington Post*'s Fact Checker have been chasing down these tales and dousing them like three-alarm fires for years. (There's even a chain e-mail that paints Snopes as a liberal cover-up for the White House.) It's often difficult for these myth-busters to say with certainty where a falsehood began."[42] The University of Indiana even has a special research project (yes, called Truthy) to monitor the viral spread of rumor to better understand the network dynamics at play.

The challenge in the age of networked knowledge and the end of Big Minds and Big Credibility is how to counter these claims without, as Dan Kahan has noted, playing into a context or worldview that would further reinforce false claims. The Obama presidential campaign responded in part by launching the Web site FightTheSmears.com, to build their own network to combat vicious online rumors. But those rumors continued to metastasize until we had Donald Trump on prime time television demanding to see the president's birth certificate. It created quite a dilemma for the president: giving in to the absurdity might look like a victory for the right, but at the same time the issue was becoming a giant distraction. In the end, President Obama made the rumor about his birth and religion a central part of the annual Gridiron roast of major political people in Washington, D.C., opening by suggesting that new footage of his birth had come to light and then showing a clip from the Disney animated feature *The Lion King*. Obama mercilessly pilloried Trump's ridiculous assertions about Obama's birth certificate—while Trump was sitting in the audience. It was humiliation by humor—preceded by the release of the actual birth certificate three days earlier, and then followed a day later with the revelation that the U.S. government had identified and killed Osama bin Laden. A brilliantly delivered three punches to put to bed the assertions that Obama's presidency was illegitimate. But even so, rumors continued to circulate online, and two years later the Arizona secretary of state went so far as to suggest Obama would not be on the ballot in Arizona unless he could prove he was born in the United States.

It's up to us to develop new ways of developing social consensus around information, so that we don't lose ourselves in a morass of opinion, hearsay, and rumor, none more vetted or reliable than the next. Considering Page's work on diversity, we're going to have a harder and harder time tackling our individual challenges if we end up in digital cul-de-sacs where we agree with everyone in our immediate vicinity. At the same time, our ability to collectively handle complex social, political, and environmental challenges will be hamstrung by increasing extremism and intolerance. If we aren't careful, we might find ourselves actually believing the rumors circulating with ever more force, falling prey to demagoguery and losing our grasp on the realities of the world.

But Wait . . . Let Me Make It Scarier for You . . .

Heck, we might find that our hard-won knowledge and technical ability becomes lost altogether, leading us back into another dark age. The system of peer review of scientific research now undergoing change bears great responsibility for the dramatic technical gains of the last century; what happens when it's gone? Will recognizable knowledge and certifiable facts evaporate entirely? If knowledge resides in networks, what happens if the networks get disrupted? Does knowledge collapse?

I can't help but think of ancient Rome. The scale of the Roman Empire was dazzling, boasting at its peak more than 50,000 miles of paved roads[43]—compared to 47,182 miles of the U.S. Interstate Highway System[44]—and before mechanical earthmovers and other equipment existed. It seemed like Roman civilization would last forever. And then it slowly disintegrated— and in its collapse knowledge and expertise that seemed essential and canonical disappeared. One of the most famous Roman innovations—their impressively resilient cement that in part allowed Roman ruins littered across Europe and the Mediterranean to endure—was lost to Europe for centuries. Fast-forward to today. Only 39% of Americans believe in the theory of evolution—a "theory" based on scientific method and underlying many of our scientific and technology advances over the last century.[45] This isn't technology's fault, but when I see broad swaths of the public rejecting established science, I begin to wonder where we're headed and whether the decline of universities and think tanks and the accompanying norms of scholarship will accelerate ignorance.

Knowledge and technical capability are so pervasive today that they seem rock solid, capable of standing the test of time. In truth, they are far more tenuous than we think, and we need institutions to safeguard them. Colin Wells tells the story of how the classics of the ancient Greek survived in Byzantine monasteries until their revival in early Renaissance Italy:

> As Latin West and Greek East [of the Roman Empire] drifted apart during the Middle Ages, Byzantine scholars painstakingly preserved the ancient Greek classics. Then, at the dawn of the Renaissance, they came to Italy and taught ancient Greek literature to the first Italian humanists, who were only then beginning to hunger for knowledge of Greco-Roman antiquity. Were it not for this small but dynamic group of Byzantine humanist teachers, ancient Greek literature might have been lost forever when the Turks conquered Constantinople in 1453.[46]

If the monasteries of the Byzantine Empire helped to preserve and protect a crucial storehouse of human knowledge during Western Europe's Dark Ages, I begin to wonder which institutions might preserve *our* most precious thinking and literature. It's almost as if authority has become so unreal in peoples' minds that "making shit up" doesn't seem like such a cardinal sin, a violation of trust, as it used to. Facts seem unreal—just more zeros and ones. Radical transparency—making as much source material and data available as possible—is an emerging new "institution" of the digital age. But transparency on its own isn't enough. In this environment, where will we safeguard knowledge? Where are the monasteries of our age? Will we even recognize them?

Authority Redux?

Here's the rub: the Internet, the same resource through which you and I may receive forwarded e-mails that purport to advance "facts" that are actually entirely fictitious, is also the greatest resource the world has ever seen for feeding the curious and discerning mind. If you're hungry for knowledge, you can dive deep into any one of thousands of topics. It's one of the things I love most about the Internet. I might hear an off-hand comment or see a footnote in something I'm reading, and, before I know it, I've spent hours reading, listening, and watching a tremendous digital archive on the subject, frequently with the assistance of Wikipedia.

As part of my class, I require my students to create a Wikipedia account and contribute to Wikipedia. As I outlined in chapter 6, approximately 5% of the world's entire population uses Wikipedia on a monthly basis. I am quite sure that a couple of decades out, my students' reputation as contributors to Wikipedia will be as important as their Harvard diplomas in credentialing their professional expertise. If you haven't tried to contribute to Wikipedia yet, you should!

I emphasize Wikipedia in this context because I interpret it broadly as a model for how we might reconstitute expertise and cultural authority in the digital age. It's not a perfect model, but at least it provides a possible path to authority and knowledge. Jimmy Wales, a cofounder of the collaboratively built online encyclopedia, describes the site's ethos as "Ignore all rules," which is further clarified to mean, "If a rule prevents you from improving or maintaining Wikipedia, ignore it." The site dramatically ignores hierarchy, keeping every decision as open to the entire community as possible. While the maintenance of such a large and robust site requires administration, the power of administrators is referred to as "mop and bucket," evoking janitorial duties rather than substantial, sweeping authority. Reputation still matters, but only in the context of your contributions to Wikipedia's quality.

Even though the site has no rules, it does boast a robust culture that revolves around two core values: verification and a neutral point of view. Contributions to Wikipedia cannot take sides; they must have a neutral point of view and provide citations for every contention or fact. Articles must rely on information from published sources—the mainstream press, published books, scholarly journals. Thus a fundamental paradox analogous to others we've seen: the world's greatest media entity, the media entity with the single greatest reach, the one that undermines the press's very authority, grounds itself in the very institutions it is disrupting. Jörmungandr, the giant snake of Norse mythology, eating his own tail into infinity, comes to mind.

In Praise of Memory

Our institutions of higher learning have notable failings. Still, for me, the decline of cultural authority and certifiable knowledge constitutes one of the most disturbing consequences of the End of Big—precisely because I suspect the old-fashioned institutions of knowledge had a unique, humanistic value all their own. I can evoke the humanist's deep, general misgivings about radical connectivity by turning briefly—and by way of conclusion—to the subject of memory. If our ability as a society to come to consensus about humans and the natural world is under siege, what about our ability as individuals to discover, know, register, and remember facts about our own lives? Gordon Bell is one of the computer pioneers who helped birth our digital age, a pivotal figure in the history of both computer processing and the Internet. Since 1998, he has been part of a great experiment at Microsoft called MyLifeBits, undertaking the mammoth task of digitally documenting every single second of his life in a searchable, archival way. He wears a video camera as well as a specialized arm strap that records his biometrics. Literally every fact about his life is recorded and measured in a digital format. Sherry Turkle describes a visit with Bell:

> [Bell] suspects his projects may be changing the nature of his memory. Bell describes a lack of curiosity about details of life that he can easily find in his [digital] life archive. And he focuses on what the archive makes easily available.... One senses a new dynamic: when you depend on the computer to remember the past, you focus on whatever past is kept on the computer. And you learn to favor whatever past is easiest to find.

The notion that technology may change the nature of memory—and consequently the nature of knowledge—has been discussed in many venues. Nick Carr wrote a memorable article for the *Atlantic* titled "Is Google Making Us Stupid?" that caused such a stir he turned it into a book, *The Shallows*. My own sense is that the technology does indeed change our memory of facts—and that we may not be more shallow, but we are certainly missing something.

My hobby for the past decade has been memorizing poems. At this point, I have somewhere in the neighborhood of three hundred poems committed to memory. They are wide-ranging, from Shakespeare and Keats to contemporary poets like Eamon Grennan and Billy Collins. My longest one is "The Love Song of J. Alfred Prufrock," by T.S. Eliot; it takes me about twelve minutes to recite the entire thing. I started memorizing poems quite by accident. In 2002, when I was living in New York, April brought National Poetry Month. I had a long commute by subway, and, one day, I finished the book I was reading, looked up, and found a snippet of poetry in the advertising space on the subway, as part of a campaign called "Poetry in Motion," inaugurated with National Poetry Month.

That shard of poetry stayed with me throughout my work that day. The next morning, I found myself thinking about it in the shower. A couple of days later, I walked past a big display in a bookstore for National Poetry Month that featured a book edited by John Hollander, *Committed to Memory: The 100 Best Poems to Memorize*. On a lark, I picked it up and spent the next two years of my commute memorizing poems. I found I loved carrying these poems with me. It is as if I walk through my life with a couple hundred poets as companions, a background chorus commenting on everything from the weather to my mood with their lines of verse. I am reminded of the George Bradley poem "Paideia," which begins:

> My poems are my children, and I swear
> on the graves of my ancestors
> I never laid a hand on them,
> not even when they exasperated me,
> when caring for them left me exhausted
> and their cries in the night disturbed my sleep.

That's how I feel about the poems I have memorized: they are my companions, my children (even if I didn't write them), and they color my every experience and provide a rich inner monologue to my life. Although I can google them at any time, carrying them around in my head has tremendous impact. I would be lonely—desperately depressed—without them. They are the birds of the John Hollander poem, chastising me for "not delighting in their brightened gray."

Rote memorization as a part of classroom instruction has mostly departed from our curriculum. And yet I find it so rewarding, so rich. I wonder—and worry—about what we may lose as our digital life, the life of radical connectivity with its persistence and its sociality, chips away at and finally dissolves the humanistic institutions of education and expertise that have sustained both our culture and our collective memory as a species. David Shapiro, in a recent essay in *Poetry Magazine,* remarked, "An analyst said I'm not demented because I still know *The Waste Land* by heart. My mother said to memorize many poems: 'It will be good for you in prison.'"[47]

T.S. Eliot, in his landmark essay "Tradition and Individual Talent," made the case for artistic genius steeped in real, clear, historical knowledge. Eliot argues that one required the other, that they were symbiotic and essential to the creation of art. I would argue that they are essential to the nature of leadership, too. As Eliot writes, "Some one said: 'The dead writers are remote from us because we know so much more than they did.' Precisely, and they are that which we know."[48]

BIG COMPANIES

We stand in the rain in a long line
waiting at Ford Highland Park. For work.
You know what work is—if you're
old enough to read this you know what
work is, although you may not do it.[1]

Katie and Chris, a friendly couple my wife and I met through our son's preschool, have an online store (SuzieAutomatic.com) that sells a product they make themselves: stunning laptop decals decorated with images ranging from the spooky (the Headless Horseman, carrying the Apple logo of my Macbook in his arms for Halloween) to the political ("This Machine Kills Facists," a homage to Woody Guthrie). Katie and Chris sketch the designs themselves on "fancy machines we call laptops" and print them on a special vinyl printer in their basement. Using Etsy, an online platform for selling things you've made yourself, they have built a solid online sales business. Tapping a company called Square, they use their iPhones to process credit card transactions when they go to fairs and craft shows. In the past, Katie and Chris would have had a hard time making a living selling laptop decals in our suburban town in Massachusetts, since the local market for laptop decals isn't large. But thanks to Etsy and Square, they've built a thriving business selling to consumers across the globe.

Katie and Chris are not alone. More Americans are starting their own small businesses—and staying small. According to data from the Census Bureau and the Kauffman Foundation, large companies shed 11 million jobs from 1997 to 2005,[2] while during roughly the same time period, from 1997 to 2008, start-ups created more than 48 million jobs.[3] Of those 48 million jobs, 31 million were occupied by the entrepreneurs themselves, while the remainder were created by a business founder for someone else.[4] That's a lot of people working in small shops. If you've got a crazy idea, a little capital, and a burning desire to build something, now is an exciting time.

Are we seeing the end of big in business? Is the huge multinational corporation on the verge of extinction? We've seen how a number of significant institutions are teetering on the brink thanks to both their own mistakes and the rise of radical connectivity. The consequences of institutional collapse have ranged from catastrophic (the loss of accountability journalism will tank our democracy) to the merely disturbing (how will our creative class, our artists, be compensated?). Through it all, I have pointed to the threat posed by the rise of Even Bigger platform companies that are helping to fuel the End of Big. The evidence in this chapter is thinner than prior chapters, but I believe it points to a fundamental challenge to scale economics and to the mega-corporations that dominate our commerce.

The end of Big Companies is at an earlier stage than the end of Big News, say, or the end of Big Political Parties—just look at the wealth currently generated by goliaths like ExxonMobil or Citibank. Yet Big Business will slowly decline over the coming decades. In the pages ahead, I'll sketch out trends in big manufacturing, big service companies, and the Even Bigger platform companies of our technological age. I'll also speculate on the moral implications of the end of Big Companies—how it will contribute to our well-being, and in fact, to civilization's continued survival.

Why Do We Have Big Companies Anyway?

It's important to remember that we haven't always had them. The corporation is a little older than 400 years, dating back to the turn of the seventeenth century, when Queen Elizabeth created the East India Trading Company. The East India Trading Company grew to become one of the largest companies in history, at one point controlling (by some estimates) more than 50% of the world's commerce.[5] It has also become something of a template for corporations: Get as big as possible, control as much of the market as possible. The East India Company's success demonstrated a fundamental tenet of microeconomics: economies of scale. Scale in business allows cost advantages and efficiency in competition, incentivizing firms towards getting bigger.

As industrialization took hold in the centuries that followed, corporations underwent several changes and shake-ups, but the pressure towards scale persisted and even intensified. Andrei Cherny, a leading economic policy maker and former senior fellow at the Harvard Kennedy School, relates that at the turn of the nineteenth century, the vast majority of workers—75%—were farmers working on their own, and many others worked in their own shops. The word "boss" didn't even enter the popular lexicon until the 1830s. "But with the rise of steam and steel, America began to creep toward industrialization and its larger, more complex economic organizations. By 1860, 40 percent of Americans worked for someone else and the economic discussion began to change. . . . By 1920, fully 87 percent of all wage earners were not only working for someone else but for a corporation."[6]

To compete in the Industrial Age, you needed lots of capital, and you also needed to sell large amounts of output to make a profit. The more you sold, the bigger you got, the more profit you made, and the more capital you could accumulate. This dynamic persisted into our own time, as evidenced by the mega-mergers that have taken place in many industries. Yet forces have also been in play in recent decades that render scale economics less powerful.

In previous chapters, I argued that internal dynamics were already bringing down big institutions, and that radical connectivity only sped up the process. With the end of Big Business, we don't see the same kind of systemic corruption necessitating the decline of big. No doubt about it, Big Companies have seen their share of scandals in recent years. In the case of the banking industry, the excessive accumulation of power by big players has short-circuited government regulation, resulting in a credit crisis that almost brought down the entire economy and from which we and the world as a whole are still recovering. Yet Big Companies are more resilient than other Big Institutions and, from a strictly functional viewpoint, probably healthier. The forces that have already counteracted traditional scale effects seem to be, more than anything else, by-products of the economy's recent evolution. If the primary element of the old economy was the industrial unit, today the Industrial Age has given way to a service economy and to what Cherny calls the Individual Age, in which the skills, talents, and labors of people matter most and people can sell their services themselves in the free market without having to work for large corporations.

Daniel Pink worked in the Clinton-Gore White House as Al Gore's chief speechwriter. In 1997, he left to work as an independent consultant or freelancer. Finding a burgeoning culture of self-employed Americans like him, he excitedly wrote a landmark article for *Fast Company* magazine about the "free agent nation" he had stumbled upon.[7] The article hit a nerve, and he went on to turn it into a best-selling book. Almost fifteen years later, *Harvard Business Review* revisited Pink's argument and found that workplace trends did indeed show an astonishing movement toward individual employment.[8] Why is that? I argue here that radical connectivity—in particular the efficiencies provided by cloud computing for sharing resources and collaboration—is dramatically reducing scale effects and will continue to reduce them in the coming decades. Add in technologies that enable on-demand fabrication, and you'll see a significant erasure of the advantages of size in commerce, and economies that are far more local than at present.

The Future That Is upon Us

I admit, talk of scale economies is abstract. Let me make the End of Big in business (or more precisely, what I speculate will be the gradual decline in Big Companies) more real for you. Within the next ten to twenty years, when you wake up and get ready for work, you will don clothes designed and sewn not by large sub-contractors for the big fashion brands, but by self-employed artisans from around the United States. You will have found your attire (which I assure you will be quite fashionable) not at a large suburban mall owned by a major real estate company but through a site like Etsy. How about shoes? Well, your brother will have recommended a shoe design to you that he found through social networks—the shoe was created by his wife's sister's roommate's nephew. You will have bought the blueprints for the shoe online and fed them to your 3-D printer. Overnight, this printer (using a technology that already exists) will have sprayed plastic into the shape provided by the shoe design blueprints. You will wake up, get dressed, open the 3-D printer (shaped almost like an oven) and take out your new shoes.

Your house will draw its electric power from a shared neighborhood renewable—wind and solar—power station, which you will have built with the help of a site like One Block Off the Grid (you can sign your neighborhood up now at 1bog.org). Your car will be an electric one that charges up at night in your garage, designed and manufactured not at a Michigan factory run by one of the Big Three automakers but in a nearby town; thanks to advances in small-run fabrication and manufacturing, there will soon be thousands and thousands of car companies, producing vehicles customized for every locale. (Think this is a pipe dream? There are already hundreds of car companies in the United States that produce small numbers of vehicles.[9])

If you're a professional, you'll work as a self-employed consultant, organized into companies of exactly one person—you—but sharing coworking spaces with about forty other people. You will have found your own workspace using sites like deskwanted.com, an online searchable database of shared workspaces. When you need a larger team for a project, you will recruit others in your coworking space or turn to the Internet to find the best talent worldwide. Members of your team will come together for a few months and, once the project is complete, disperse and turn to other projects.

You get the idea. If today, bloggers can publish anything at any time to any audience at zero cost, within the next twenty years everyone will enjoy the capability to be their own Walmart. As radical connectivity continues to advance, and as it increasingly comes to affect fabrication and manufacturing, anyone will be able to design and sell anything, and anyone else will be able to buy anything. That's right—*anything!*

The Collapse of Scale

Now, I can hear the skeptics: Will everyday individuals really achieve Walmart's capacity to sell goods globally? Maxwell Wessel seems to think so. He's part of the Forum for Growth and Innovation, a Harvard Business School think tank developing and refining theory around disruptive innovation (innovations that disrupt existing markets). Wessel argues that scale, "one of the last bastions from the competitive storm," is no longer profitable or safe. For a long time, technology made being big very profitable and very safe—because no one else could afford to be big. If you were Ford Motor Company, no one else could afford the vast machinery needed to manufacture cars. If you were Walmart, no one else could afford the complicated technology to track the demand logistics from shoppers in Peoria to production plants in China. But radical connectivity allows small businesses to collaborate in loose ways that give them capacities comparable to those of Walmart or Ford Motor Company. "If you want to sell stuff made in China, you no longer have to have representatives on the ground in that country, nor do you have to place orders for a certain, guaranteed volume of goods. All you have to do is click on a site like Alibaba .com." As Wessel points out, other companies now exist that provide twenty-four-hour call centers and state-of-the-art business management software to small organizations. "The competitive advantages of scale are being commoditized. Minimum efficient scale is getting smaller and smaller.[10]

The impact of the collapse of scale will change—in fact, already is changing—our economy and our companies in ways we do not yet understand. In some ways, this is a natural next step in the recent history of manufacturing. During the twentieth century, manufacturing required a lot of direct workers, all located here in the United States. Then manufacturing moved overseas and became increasingly automated. As it gets more automated, the costs of smaller manufacturing runs has dropped, to the point where we can now make one of something economically. Radical connectivity has played a key role in collapsing the advantages of scale, removing barriers to entry to the marketplace and—more importantly—allowing small companies to share resources that previously were only available to Big Companies.

Inside the Cloud

Cloud computing exemplifies such resource sharing. You hear people talking about the cloud all the time in the context of the Internet, but a lot of us still have a fairly cloudy notion of what exactly cloud computing means. From the perspective of companies, cloud computing makes it easier to share computing resources that otherwise would be too expensive. Let's take a look at how this works.

The Internet physically exists on computers called servers. Technically, any computer can be a server, even your smartphone. (For a while, I kept a server in my closet to host my personal Web site, http://nicco.org, using an old laptop I had retired from travel duty.) But these days, most servers are specialized computers, designed for hosting Web sites. Large Internet companies like Google and Facebook invest billions in server farms—large facilities with rows upon rows of computers that physically host the Internet.[11] Server farms generally require a large amount of electricity, not just for the computers but for the cooling systems required to keep the temperature down (ever notice how hot your lap gets when your computer sits on it for too long?). Companies that require large server farms are always looking for sites near cheap sources of power, sometimes through price breaks or tax incentives that lead to local political issues.

Of course, when you're surfing the Internet, you don't care where the Web sites physically reside. You're in the virtual cloud of the Internet, and the specific server—be it in Idaho, New York, or Shanghai—doesn't affect your experience. Indeed, servers have become incredibly commoditized, with large volumes of computing power made available in seconds for pennies. Amazon has developed some notoriety in this area with a product called Amazon web services (AWS). In the process of building a giant infrastructure to sell everything, but especially books, over the Internet, Amazon realized that they could sell excess capacity on their server farms. Need a place to host your Web site? Buy it from Amazon. Suddenly need 10,000 times more space because your tiny start-up has gone viral? Amazon can turn it on in seconds.

Hundreds if not thousands of vendors now offer cloud computing. The complicated technical resources required to host and manage Web sites used to cost a lot of money and take a lot of time. They still do. But thousands or even millions of small businesses, groups, or individuals can now share that cost, allowing your tiny start-up access to the same kind of computing resources mobilized by a giant corporation like Google or Facebook. This kind of level playing field offered by cloud computing threatens Big Companies—and it ultimately even challenges the hegemony of the large, Even Bigger platforms described in chapter 4. Harvard Business School's Wessel articulates, in effect, an argument for why loose coalitions of small businesses banding together can achieve comparable competitive power as large platform players like Amazon.

EchoDitto In, Dewey LeBoeuf Out

Take my company, EchoDitto. We're a consulting firm with twenty-five employees, doing both technical and strategic consulting. In the past, if we wanted project management or time management software to optimize our service consulting business, we would have had to hire a bunch of programmers to design and build the software. Or perhaps we would have bought an expensive software platform and had it customized and installed on a server in our office. Today, thanks to the cloud, we can go online and buy what amounts to a share in any one of a range of impressive, complicated software products, obtaining a competitive edge that in the past only a much bigger firm would have enjoyed.

Service consulting businesses of all kinds are today rapidly adapting to the End of Big. A great illustration is the 2012 collapse of Dewey & LeBoeuf, one of the oldest and largest law firms in the United States. Founded in 1909, the firm employed more than 1,400 lawyers at its height. A major financial scandal at the top of the firm hastened its demise, but that's not the full story. Stuart Saft, a former partner in the law firm, had this to say: "[In the past] law firms were partnerships. It was an institutional practice, not individuals with portable books of business. Now, everyone has become a free agent. It has changed and destabilized the nature of the legal profession. And once it takes hold, it accelerates. Other firms say they don't do it, but I wonder if that's really true."[12]

In the past, to have big and powerful clients as a lawyer, you needed to be at a big firm, because only a big firm had access to a wide range of resources: large research teams, expensive subscriptions to legal journals, secretaries to help you manage large volumes of paper that provided you with a research advantage. Today, with the help of simple online services like Google Search and specialized cloud-based software for the legal profession, a big firm's allure has substantially diminished. "If you're young and starting out, a big firm still gets you entrée into the elite social networks that lead to clients who pay large fees," a former junior lawyer from a big firm tells me, "but other than that, they don't provide much of an advantage. There used to be the opportunity to learn from more senior, experienced lawyers, but even that culture has disappeared."

In their 2006 book *Revolutionary Wealth*, futurists Alvin and Heidi Toffler imagined work without firms, a world where everyone is a freelancer. When you had a project that needed doing, you'd collect the best bunch of freelance experts to do the job, and when the project was done, the opportunity seized, you'd disperse. As Daniel Pink demonstrated, that day was already starting to arrive during the 1990s. Today, it's upon us in full force. The number of self-employed people who are essentially one-person consulting shops has skyrocketed since 2001;[13] by some estimates, more than 42 million people work part-time or on their own—more than the total number of autoworkers, teachers, and doctors combined.[14] This seismic change in the economy is not without its pressures. Gone are the days of cradle-to-grave employment. How much of your time do you want to spend worrying about where your next project—and paycheck—will come from?

Dawn of the Replicator

If cloud computing has brought the End of Big to service businesses like law firms, 3-D printing and on-demand fabrication will do the same thing to companies in manufacturing-based industries. In my office, I have an inking book I found at a garage sale for $1. It's from 1942, and it contains page after page of different typefaces. The idea was that you would trace a typeface from the book, and then ink it to make a poster or headline. Hand-drawing fonts seems like a quaint throwback these days. In the late 1970s, the term "desktop publishing" started circulating. It was a crazy idea at the time—the notion that anyone could design something on a personal computer and then print it right there, full-color photos and all. But by the mid-1990s, color desktop printers were common.

We've lived this revolution together. Remember what it used to be like to take a picture? As one blogger recounts, "Just twenty years ago, taking a picture meant using a mechanical camera to expose film to light, baths of toxic chemicals and a darkroom to develop that film and make prints; editing those photos involved tools for cropping and the laborious use of airbrushes and still more chemicals. Now, with a camera the size of a pack of cigarettes, a cheap computer and printer, and some free software, anyone can shoot, doctor, and print photos in a matter of minutes."[15]

The same immediacy and personal power that color home printers have brought to photography and publishing is coming to physical things. While your desktop printer will print text and images, a 3-D printer creates actual stuff, objects you can touch. Different printers do this differently, but one of the cheapest 3-D printers available is called a MakerBot and runs about $1,850. You plug your computer into the printer, send it the designs for whatever you want to print, and it will spray melted plastic together to create your design, as long as it's smaller than a grapefruit.

Sound crazy? Go buy a MakerBot and try it. I did. It's nuts. The first thing I printed was a replacement train track for my boys. They're four and two and obsessed with Thomas the Tank Engine. The bridge for their train set had broken a part of the track, so I downloaded the blueprints from the Internet, fed it to my MakerBot, and about thirty minutes later we had a new track piece to make the bridge operational again. When I spoke with the MakerBot founder, Bre Pettis, I asked about a future where everyone can print everything at home. What's going to happen to big companies? "I'm not convinced that giving everyone production in their house is the best use of their time and money," he said. "Printing a garden chair might take a week versus buying it at Walmart." Still, he pointed out that the ability to print anything is pretty huge, "empower[ing] people to think about alternatives, about how things are made."

His customers have been not just thinking but printing out all kinds of things. You can buy a MakerBot at http://store.maker bot.com/, but once you've got one you need blueprints of things to print. I'm not skilled at design, but many other people are. They've uploaded more than 15,000 designs of things you can print to Thingiverse, a Web site maintained by the people who sell the MakerBot. A quick review of recent Thingiverse uploads shows a replacement flipper for a vintage pinball machine, latch for a a guinea pig cage ("made to lock a cage that we received without any way to lock the base to the upper part"), a replacement knob for a stove top, and a replacement part for a broken blender. The *New York Times* has profiled an ingenious invention for sugar-crazed kids everywhere: the "Lucky Charms Cereal Sifter." This essential device allows you to separate out the much-coveted marshmallows from the less tasty cereal bits. No sooner had its path-breaking design (the user shakes the sifter, causing the cereal to plunge to the bottom but keeping the marshmallows) been uploaded onto Thingiverse than another entrepreneur saw fit to offer a fully assembled version on another site for $30.[16]

The implications of the MakerBot and similar technology are hard to fathom. "My grandfather saw the advent of radio," says Pettis. "My dad witnessed the advent of the TV. Then, in my lifetime, we've had the computer, the Internet, and more and more and more. . . . Technology is taking off. Who knows what's possible?"[17] As Steve Denning has written in a recent issue of *Forbes*, 3-D printing will revolutionize manufacturing, rendering it far more local and smaller scale: "Now the economics of large-scale production runs carried out overseas are being undermined by the possibility of making, selling and delivering millions of manufactured items one unit at a time, right next to the customer. Digital manufacturing is beginning to do to manufacturing what the Internet has done to information-based goods and services. . . . [A] massive transition from centralized production to a 'maker culture' of dispersed manufacturing innovation is under way today."[18]

Today you have to spend about $2,000 and some time and you've got your own 3-D printer. But it's easy to imagine that ten years from now, every home will have a 3-D printer, just like every home today has a microwave. Over time, these 3-D printers will grow more advanced. Once nanotechnology hits its stride, 3-D printers will build complex machines such as an iPhone in your own home. In fact, Apple is already preparing for this future. Take the iPhone you've got in your pocket right now and turn it over. You'll see that it says in fine print, "Designed by Apple in California. Assembled in China." Apple understands that the design is the important part of what companies do, not manufacturing, and it is already staking its claim to the design.

The Coming Plague of *Shanzhai*

On-demand fabrication has the potential to do a lot of good around the world. A group of smarties at MIT has developed the Fab Lab—a kit of about $20,000 worth of equipment that allows all kinds of things to be manufactured on-demand. I mentioned the fab lab in Afghanistan earlier. These labs also exist in Boston, Norway, India, Costa Rica, and Ghana, focusing on things needed by local communities, such as wireless networks, educational tools, and agricultural instruments. In Ghana, for instance, the lab works with local users to create solar-powered devices that help with cutting, cooling, and cooking so as to meet local needs.[19] The science fiction writer Bruce Sterling imagines how fabrication can also assist with disaster relief and refugee camps: "A big, bad, cheap fabricator that makes stuff out of utterly worthless raw materials. Straw and mud, perhaps. Or chopped grass, cellulose, recycled plastic and newspaper, even sand. . . . You need buckets. The mobject-maker spits out these general issue buckets. Khaki-colored maybe, the color of mixed dirt. Ugliest buckets in the world, but they work. They carry water. Now you need latrines, so out come a few hundred of them. Sewer pipes. Shower stalls. Faucets."[20]

Yet the implications of 3-D printing aren't all positive. While the last decade has opened all forms of media to piracy, the next decade will open up everything else—all property. See something at a friend's house you covet? Take a photo of it with your phone. Go home, feed it into your computer, and send it to the 3-D printer. A little while later—voilà! You've got your own copy! "You know the machine on *Star Trek*? The replicator? That's what I was aiming for," says Jim Lewis of eMachineShop.com.[21] You can go online right now to Jim's Web site, design just about anything you want—a new coffee table, a pair of shoes, or (like the *Wired* reporter Clive Thompson) an electric guitar, and eMachineShop will print it for you. With MakerBot, the objects have to stay small and plastic, but on eMachineShop you have enormous flexibility and choice of both size and material.

Widespread undermining of property rights thanks to shifts in manufacturing is already happening. Across major manufacturing regions in China, *shanzhai*, or "bandit" manufacturers, are mashing together different brands and designs to create new objects in short-run fabricators that resemble industrial-scale MakerBots. Many of the objects created hold specific appeal in Chinese culture—like "traditional Chinese slippers with a Nike swoosh or Adidas stripes . . . a mobile phone that can store your cigarettes and light them for you too . . . phones shaped like beetles, pandas, or Mickey Mouse."[22] As the blogger Bunnie Huang writes, the *shanzhai* "are doing to hardware what the web did for rip/mix/burn or mashup compilations. . . . They are not copies of any single idea but they mix IP from multiple sources to create a new heterogeneous composition, such that the original source material is still distinctly recognizable in the final product. Also, like many Web mash-ups, the final result might seem nonsensical to a mass market (like the Ferrari phone) but extremely relevant to a select long-tail market."[23]

The Power of Quirky

Apple might think that its emphasis on design will enable it to protect its position, but in fact radical connectivity—and, specifically, the crowdsourcing it enables—is beginning to commoditize even design and other forms of intellectual property, reducing the commercial impact of scale. At night, I plug my laptop in and leave it on my dresser (high enough that little hands won't find it) to charge. But in the morning, when I unplug the laptop, the power supply always slides behind the dresser. I got sick of beginning each morning with a tinge of frustration, so I started looking for solutions. One day, while I was at Target, I noticed a perfect, ingenious little solution: a paperweight-cum-cable-holder to solve exactly this problem. I bought one and took it home. It worked so well I went back and bought three more. When I opened it, a little pamphlet fell out, and there were the names of the 562 people who helped to design this product on a Web site called Quirky.com.

Got an idea for a product? Visit Quirky.com, pay a $10 registration fee, and share the idea. The online community on Quirky will discuss your idea, finally voting on it. Once a week, the community chooses an idea to develop. The online community works together to figure out how to get it built and how much to sell it for. If the product gets enough preorders to be profitable, then it actually goes to production and ends up on sale online and in top retailers like Target, Toys"Я"Us, and Bed Bath & Beyond. If your idea is chosen, you share 30% of the resulting revenue with the other members of the community who helped design the product. Many of the products the company ends up making fill odd little gaps. In addition to my paperweight–cable holder, there are a wide range of cases and other accessories for iPads, mobile phones, and laptops—as well as more interesting products like a portable nightlight for children and a fascinating set of cooking tools.

Quirky is a crowdsourcing solution, harnessing the individual power of radical connectivity to design innovative products. "Quirky can make anything," the company founder, Ben Kaufman, told me in a June 2012 interview. "Our only constraint is reach: advertising to reach more people." I asked him about the Big Companies he has teamed up with to distribute Quirky products, and he pointed out that sale of Quirky products through big retail chains was critical for expanding reach and establishing his products in consumers' eyes, but he could also do well just by selling through the Quirky Web site. "I don't need the big logos, but I like them—they add legitimacy to our work, to our community."

Like them or not, Kaufman may have to do without them. Research suggests that the end of Big Brands may also be near. Radical connectivity has made individuals so powerful that they come to their shopping armed with a range of data and opinions from both experts and their social networks, not to mention scores and scores of online reviewers. Consequently, consumers are less loyal. As *Wired* magazine reported, one study found that "nearly half of those who described themselves as highly loyal to a brand were no longer loyal a year later. Even seemingly strong names rarely translate into much power at the cash register." In another study, "just 4 percent of consumers would be willing to stick with a brand if its competitors offered better value for the same price.[24] It turns out that the more technology you use, the more likely you are to have a lot less brand loyalty.[25] From Groupon to iPad shopping apps to Amazon, consumers respond overwhelmingly to price and to the recommendations of trusted peers in their social networks—not to brands.

Crowdsourcing works well for designing large, complex, expensive products purchased by organizations, not just small consumer items. The U.S. military has begun designing its vehicles using crowdsourcing—with some initial promise. In four months, a team designed, built, and deployed an "experimental crowd-derived combat support vehicle" or XC2V for short. The site used to design it is called LocalMotors, which "harnesses the creativity of the world's underemployed car designers."[26] Thousands of people on the site participate in the design of vehicles, some of which actually find their way into production. You can reserve your limited edition Rally Fighter—a racing car collaboratively designed by almost 3,000 people—at http://rallyfighter.com/ for $74,900. Did I mention you have to assemble the car yourself? The price includes a six-night hotel stay and "an expert Builder Trainer" to help you put the vehicle together.

Or perhaps you want a television advertisement for your company but don't have the money to pay a big Madison Avenue firm. Post your creative brief on GeniusRocket.com, a site I cofounded, and offer a cash reward. Maybe you make a new bottled soft drink, and you know it sells with dog lovers. You post on GenuisRocket.com a description of the soft drink and then explain that the TV ad must include a dog. Depending on the size of the reward, you might have a couple hundred people upload short videos they've made to compete for the reward. You choose the best one, and the creator wins a cash prize—as much as $30,000. That might sound like a lot, but, believe me, it's still a whole lot cheaper than an expensive ad firm. It's part of the "free agent" world, where radical connectivity and cloud computing allow individuals to operate outside the bounds of a normal nine-to-five job at a big company—with all the uncertainty and opportunity that comes with freelancing.

How far can we take crowdsourcing? "Giffgaff" is a Scottish English word that means "mutual giving," which is a pretty good way to describe the British mobile phone company giffgaff. Its customers are its sales team. Its customers are its technical support team. In fact, giffgaff uses customers to do as much as possible, compensating them with a virtual currency, "payback," that can be redeemed for mobile phone minutes, cash, and other rewards. By distributing key expenses out to the "crowd"—in this case, the customers—this mobile phone network in the U.K. keeps costs down and establishes itself as a competitive choice.

Open Source Takes on the Big Guys

Crowdsourcing has its roots in open-source programming, which first took hold during the 1960s. Open source started as the "free software movement," but over time it has come to encompass much more than just software. When you call something "open source," you're saying that every part of it is available to people, that the design and production is literally open for anyone who wants to take a look and maybe try their hand at improving it. Bre Pettis publishes the blueprints for his 3-D printer, MakerBot, so that anyone who has the parts can build one—or better yet, so that people can try their hand at improving the design of the printer. If you don't like how something works, open it up and change it—and then share your improvements with the world. That's the ethos of open source.

Not everyone is going to open things up and change them or improve them. Most people may not care; they want a computer that works so they can send their e-mail, and they're not going to take the time to learn how to improve it. But open source proceeds on the theory that "given enough eyeballs," everything gets improved. You don't need everyone to work on improving an idea; just a small number of people working on it can make it a lot better for everyone. Open source is a powerful philosophy and a leading force behind the End of Big—and it can provide a template for building successful institutions in the age of radical connectivity. Some of the most complicated technical projects in the world—not to mention more than a few successful companies—are built around an open-source process and community.

Open-source software generally has a founder who starts by writing some original snippet of code—what we nerds call "scratching your own itch." In most successful open-source projects, a leader or group of leaders champions the software's use and encourages people to improve it, sometimes setting directions for ongoing development. Yet despite the presence of leaders, the community is paramount, because the value in the project resides in the community, not in the founder or leader. The community builds the software, and the community uses the software, whereas the leader provides direction and inspiration and occasionally resolves community disputes.

Even though developers of open-source code give it away for free, many successful businesses emerge on the back of open-source ideas—they just tend to be small. You could search out all the parts to build a MakerBot by yourself, but it's easier to just buy all the parts in a single kit. If you use OpenOffice, the open-source competitor to Microsoft Office, you can buy technical support for $30 a year. I've built EchoDitto, a successful small consulting company, on top of open-source software like Drupal, a content-management system for building Web sites.

Microsoft has found itself fighting open-source software on multiple fronts. For a short while, it looked like Microsoft defeated Netscape, and that Microsoft Internet Explorer would dominate Web browsing. But people didn't like Internet Explorer; it could be slow, it crashed a lot, and there were concerns about Microsoft's ability to control the Web-browsing experience. In 2002, a sixteen-year-old in Florida named Blake Ross and a twenty-one-year-old New Zealander named Ben Goodger decided to build a competitor. For three years they worked—unpaid—with other programmers in the open-source community, and, finally, in 2005, Firefox was available—free— to anyone who wanted to use it. In a matter of months, Firefox decimated Microsoft's market share in Web browsing. Ross went on to start Spread Firefox, an all-volunteer online community whose sole purpose was to convince people to switch to Firefox. As more and more people started using Firefox, the Spread Firefox community got more and more fired up. They raised money to run an ad in the *New York Times* to encourage people to switch to Firefox. More than 10,000 people donated an average of $25 each over the course of about ten days. On

December 16, 2005, the *New York Times* included a two-page spread advertising the Firefox browser—with the name of every single one of the 10,000 donors who funded the ad in tiny type.[27] Internet Explorer lost dramatic market share to Firefox, and eventually lost market dominance to Firefox and a host of other browsers like Google Chrome.[28]

Think of how incredible it is: One of the largest—and wealthiest—companies in the world (Microsoft) lost its market dominance to a free product made by a bunch of volunteers, led by a nineteen-year-old with a marketing budget funded by a large number of other volunteers. Firefox is only the tip of the iceberg; almost every major software product out there has a significant open-source competitor, with lots of small firms— like mine—helping to implement and maintain that software for other companies.

The Necessities of Life

It should be clear by now how 3-D printing, crowdsourcing, and open sourcing are bringing the End of Big to service companies (like law firms), manufacturing companies, and technology companies. But it hardly ends there. Virtually every sector of the economy is and will be affected, including some of the most necessary and fundamental for our continued prosperity. Take farming. As the noted activist and intellectual Bill McKibben has informed us, the long-standing decline of family farms in the United States has been reversed over the past couple of years, driven by the reemergence of small farms producing food to supply local demand. These farms "are not yet a threat to the profits of the Cargills and the ADMs, but you can see the emerging structure of a new agriculture composed of

CSAs and farmers' markets, with fewer middlemen. Which is all for the good. Such farming uses less energy and produces better food; it's easier on the land; it offers rural communities a way out of terminal decline . . . this will be a nimbler, more diversified, sturdier agriculture."[29] Yet small farming is more than an "emerging structure"; in 2012, BusinessWeek reported that "after 18 years of steady increases, the number of farmer's markets across the country now registered with the USDA is 7,864. In 1994, there were 1,744." According to the USDA, that represents an estimated $1 billion of transactions last year alone.[30]

A few years ago, a group of friends in North Carolina started something they called a Crop Mob. Over the Internet, they organized a fairly large group of people—approximately 150—and visited a small local farm to help with the October harvest. In a few hours, this Crop Mob did literally weeks' worth of work, harvesting more than 1,600 pounds of sweet potatoes and helping the farm remain profitable. Many small farms that operate on the margins desperately need the extra labor force, and Crop Mobs can be tremendously helpful. Word of the North Carolina Crop Mob spread online, and a year later more than sixty popped up across the United States to help small farms with labor-intensive tasks, including one in Seattle that works on community gardens.[31] Crop Mobs have given rise to other projects, like the Food Matchit Project that is designing online systems to help support local agriculture by knitting together labor, tools, land, transportation, processing facilities, points of sale, knowledge—even compost!

Such initiatives in farming are reassuring, because given the growing specter of climate change and rising populations (currently seven billion humans on the planet, and we're not so far away from eight billion), our basic systems for providing the necessities of everyday life need to change. David Crane, president and CEO of one of the largest power suppliers in the U.S., NRG Energy, Inc., sees the End of Big coming to energy. As he remarked during a conference call, "we will be in a situation where within two years the price of delivered power from solar installations will be able to undercut the retail price of grid power in roughly twenty states. . . . This low-cost solar power, installed in ever-increasing volumes on a distributed and semi-distributed basis in a way that obviates the need for a lot of very long high-voltage transmission lines, has a potential to revolutionize the hub-and-spoke power system which currently makes up the American power industry."[32] Right now you get your power from a fossil-fuel power plant owned by a big utility or power company. But your neighborhood could set up a shared solar installation to create power for nearby homes (using the blueprints and guidance of the crowd-sourced online community SolaRoof.org), and could even conceivably sell that power to other nearby neighborhoods—puncturing the model of Big Power.

The financial backbone of the economy is also experiencing the End of Big. Big banks are (even as we speak) doing everything they can to resist being broken up into smaller entities, but they're losing ground to small local banks as well as online start-ups that offer alternatives. In a single month, September 2011, approximately 650,000 customers, shifted more than $4.5 billion from big banks into the nation's roughly 7,000 credit unions according to the Credit Union National Association (CUNA), in part in reaction to Bank of America's $5 debit-card fee (which was later conceded).[33] As big banks continue to find ways to fund their scale, smaller banks will have more and more opportunities to offer.

So, too, will nontraditional players. Square, a start-up founded

by the Twitter cofounder Jack Dorsey, sells a little piece of plastic that plugs into the headphone jack of your iPhone or An-
droid, allowing you to swipe credit cards. In effect, Square turns
any smartphone into a way to charge credit cards, allowing anyone to have the merchant services previously afforded only to
small businesses that met a certain threshold of activity. And
the best part about Square is that it's free, although you will pay
a 2.7% transaction processing fee on every credit card you swipe.

Small Craftspeople Will Rule the Earth

Ultimately, the collection of trends that comprise the End of
Big in business will, at its best, lead to the rise of a craft-centric
economy. We're seeing some of this already, with the rise of
Etsy.com and the so-called maker culture. Maker culture is not
just about crafts; it is about things like solar roofs, 3-D printers,
and nearly anything else you can think of. But the craft aspect
stands out and harkens back to a return to a village way of
thinking about the world. As I mentioned at the opening of this
chapter, my friends Katie and Chris sell on Etsy a product they
create themselves. To date, Etsy has more than 12,500 online
stores featuring the work of thousands of makers. Etsy represents a new kind of small business, and a new approach to
commerce. People are turning away from big companies in favor of stuff created by real live human beings that they can get
to know. It's a return to an old kind of value system, one rooted
in American self-sufficiency and innovation. But it also offers
some clear drawbacks. For one, producing things by hand is a
lot of work. Yokoo Gibraan sells hand-knit scarves on Etsy; she
makes more than $140,000 a year. But as she tells the *New York
Times*, it comes at a cost:

"I have to wake up around 8, get coffee or tea, and knit for hours and hours and hours and hours," said Ms. Gibran. "I'm like an old lady in a chair, catching up on podcasts, watching old Hitchcock shows. I will do it for 13 hours a day." And even after all those hours knitting, she is constantly sketching new designs or trading e-mail messages with 50 or more customers a day.[34]

I spoke with Maria Thomas, the former CEO of Etsy. She agrees that the promise of Etsy was to encourage "local living economies," but she also noted the critical role of the global supply chain for most of the small businesses to be successful. Knitting scarves is possible because the wool might be grown on sheep in Lubbock, Texas, but shipped to China for processing and dye, and then shipped back to New York to be sold in a store.[35] But even so, the rise of crafting requires businesses to stay small. If you're knitting each scarf by hand, you're not going to be able to sell an infinite number of them.

There is something exciting, even thrilling about the end of Big Business. Opportunities abound for everyone, and the possibilities for innovation are fantastic. But let's take a moment to consider what gets lost. We've already seen the limitations of craft—it requires an enormous amount of work without the support of traditional employment. Another concern has to do with the reliability provided by Big Companies. Capitalism, in the sense of substantial accretions of capital that underpin economic activity, has produced big things such as airplanes and skyscrapers. When I'm flying across the Pacific, I have some comfort knowing that a large company such as Boeing is capable of hiring the very best, MIT-trained engineers to design and build their airplanes and that big government has regulations at the FAA double-checking Boeing's decisions. It

makes me feel better; I know I'm safe because of an intercon-
necting web of institutions that operate to encourage the best
and make sure my interests are protected. In a crowd-sourced,
maker world, we're missing some of the institutions that have
helped to provide accountability and reliability.

Just as the end of Big News has meant a demonstrable loss
of accountability journalism, we might worry that the end of
Big Companies could mean a demonstrable loss of accountability, especially in industries where technical expertise or competency is required. While I tend to sympathize with the views
of Ben Kaufman of Quirky and others who believe that
anything—literally anything—can be crowd-sourced, I'm not
sure everything *should* be crowd-sourced. Would you want to
fly on a small airline that uses a craft- or maker-produced airplane whose engines come from a craft- or maker-produced
engine company? Would you want to buy prescription drugs to
treat cancer from a one-person company, a chemist in his
basement concocting treatments? Would you really feel com-
fortable buying your car from a maker company you'd never
heard of before? Hey, the Chevy Nova used to explode because
of a gas tank problem. How will you know that the one you
bought from "Chip's Cars" won't?

When you couple the end of big in business with the end of
big in journalism and government, you realize we're headed
for a serious accountability gap. In this respect, we might even
return to the nineteenth century, when any quack could make
bizarre advertising claims, and you had no way of knowing if
you were getting medicine or baking soda. As with other areas
of life, it will be up to us in the years ahead to design new institutions and mechanisms to allow us to compensate for the loss
of Big Companies and create an economy that safeguards both
social order and the values we hold dear.

Toward a Better Future

It's hard to know what the end of Big Companies will mean for us. One thing is clear: Despite the power of craft and the do-it-yourself attitude that is emerging, big companies won't vanish overnight. Companies like Microsoft have enormous cash reserves that they can use to keep the End of Big at bay for a while. But smaller companies will be able to out-compete larger companies, reducing any advantage afforded by size. Consumers will have to develop ways of distinguishing between good, high-quality designs and poor ones—a function that big brands have traditionally served for us. As Bill McKibben says, "We're moving, if we're lucky, from the world of few and big to the world of small and many. We'll either head there purposefully or we'll be dragged kicking, but we've reached one of those moments when tides reverse."[36]

In some of the previous chapters, I've greeted the End of Big with considerable trepidation and even outright fear. By comparison, I regard the End of Big in business as, on the whole, an extremely promising development. Although fragmentation will create new headaches, making it more difficult for government to regulate and assure quality and safety, consumers and society as a whole will benefit from the great dynamism and innovation of small, local enterprises. I would go further and argue that the End of Big in business represents one of the greatest hopes for saving our civilization from the environmental dangers that threaten to sink it. Our current, big economy is unsustainable—almost everyone in business realizes that today. A more fragmented economy comprising primarily small, dynamic firms has the potential to lead us to more sustainability while also fostering community and continued wealth generation.

I've already mentioned how radical connectivity is on the cusp of revolutionizing farming and energy production. These two developments alone could go a long way toward helping us overcome a primary feature of our global economy that renders it unsustainable: our dependence on fossil fuels. Our global commerce is built on cheap, available oil. I can buy grapes from South America in my supermarket in Massachusetts for a few dollars because it is astonishingly cheap to ship the grapes over approximately 5,000 miles. But not for long. Christopher Steiner, a senior reporter at Forbes, has estimated that if (when?) gas hits $14 per gallon, WalMart (the world's largest truck fleet) will not be able to afford to ship goods from China and other manufacturing centers to their big box stores.[37] Paying $14 a gallon for gas may not be so far off; as the noted author and journalist Peter Maas has noted, "even the oil companies themselves—who are the most optimistic—say 'maybe another 20–30 years, we'll be able to maintain or increase the amount of oil we're producing every day. . . . That is the tail end. That is the twilight."[38]

Moving to small in food and energy production helps free us from oil in part by allowing for more efficient local distribution. But the End of Big also promises to make our economy more sustainable in other important ways. In developed economies, excessive consumption has levied a huge toll on environmental resources. Radical connectivity is coming to the rescue, fueling an emerging trend that could reduce over time how many resources we use: shared or small consumption.

Your car spends a lot of its time parked, not being used. What if you could rent out all the time your car spends idle? The Web makes that possible, and a number of companies are out there to do exactly that; RelayRides.com, for example, allows consumers to "rent cars from people in your community." Unlike traditional car rental companies, platforms like Relay-Rides, Zimride, Spride, and Getaround don't actually own any cars; rather, they provide a suite of tools—reputation ratings, scheduling tools, payment systems—that let you share your own car. The result is a far more efficient use of resources, as well as a greater sense of community that comes from sharing. As two observers have remarked, "The convergence of social networks, a renewed belief in the importance of community, pressing environmental concerns, and cost consciousness are moving us away from the old top-heavy, centralized, and con-trolled forms of consumerism toward one of sharing, aggrega-tion, openness, and cooperation."[39] Lisa Gansky, one of the pioneer entrepreneurs of the digital eye calls this new economy "the Mesh" because of its networked enabled efficiency.

Collaborative consumption might sound idealistic, but it isn't a passing fad. According to *Fast Company,* a whole "new genera-tion of businesses" is emerging that enables "the sharing of cars, clothes, couches, apartments, tools, meals, and even skills. The basic characteristic of these you-name-it sharing market-places is that they extract value out of the stuff we already have."[40] Three million people so far have used shared con-sumption sites to rent space on peoples' couches, and over two million have used them to share bikes. More than $2 billion worth of goods and services have been exchanged, without money, on Bartercard.com. Freecycle has 5.7 million members across 85 countries. (Once, while working on a political cam-paign and short on cash, I furnished an entire apartment com-plete with refrigerator and washing machine from Freecycle.) By 2015, more than 10 million people in the United States and Europe will belong to a car-sharing service like Zipcar.[41] AirBnB, one of several Web sites that allow people to share their empty guest bedrooms with strangers, now lists more available rooms in New York City than the largest hotel in town.

Our economy and our civilization are at a critical juncture. As Bill McKibben related in *Rolling Stone* magazine: "June [2012] broke or tied 3,215 high-temperature records across the United States. That followed the warmest May on record for the Northern Hemisphere—the 327th consecutive month in which the temperature of the entire globe exceeded the 20th-century average, the odds of which occurring by simple chance were $3.7 \times 10\text{-}99$, a number considerably larger than the number of stars in the universe."[42] We have a moral imperative to move away from oil, and radical connectivity can unlock a range of opportunity in the small—from sharing to more vibrant local economies. Most of our modern institutions were designed from a lens of scarcity rather than abundance. We need to start thinking about alternative ways of organizing our economies, and our work, to acknowledge and take advantage of technology. Failure to act will in all likelihood result in environmental conditions that, as McKibben asserts, are straight out of science fiction. Yet the End of Big in business can create another future we can only barely grasp today, one potentially much closer to our dreams. It's this future that I encourage you to embrace—and to help imagine.

BIG OPPORTUNITIES?

Transfiguration. Consider it from where you stand.
Overnight the cold, cloudy wet spell was lifted, and
you wake beneath a Byzantine blue dome of glass[1]

A modest ritual my wife and I have is to go see a movie together the afternoon of Christmas Eve. With small children, we don't get out much for movies, and usually it has been the only movie we see in the theater all year long. In 2011, we went to see the Oscar-winning *The King's Speech*, a film I found compelling given my own struggle with a stutter. Just watching the opening scene, which shows King George VI struggling to give a speech in a giant stadium, gave me a panic attack.

After the film was over, I found myself wondering about that epoch in Western history. I grew up in Asia, so my knowledge of early-twentieth-century European history was a bit thin. Ever the history buff, I started to consume books about the period. Barbara Tuchman's *The Guns of August* is exceptional; it opens with the funeral of King Edward VII (the grandfather of King George VI, whose story is told in *The King's Speech*). This is the book's very first paragraph:

> So gorgeous was the spectacle on the May morning of 1910 when nine kings rode in the funeral of Edward VII of England that the crowd, waiting in hushed and black-clad awe, could not keep back gasps of admiration. In scarlet and blue and green and purple, three by three the sovereigns rode through the palace gates, with plumed helmets, gold braid, crimson sashes, and jeweled orders flashing in the sun. After them came five heirs apparent, forty more imperial or royal highnesses, seven queens—four dowager and three regnant—and a scattering of special ambassadors from uncrowned countries. Together they represented seventy nations in the greatest assemblage of royalty and rank ever gathered in one place and, of its kind, the last. The muffled tongue of Big Ben tolled nine by the clock as the cortege left the palace, but on history's clock it was sunset, and the sun of the old world was setting in a dying blaze of splendor never to be seen again.[2]

It was an astonishing show of pomp and circumstance, and, to contemporaries, it evoked the undeniable power and stability of monarchies across Europe and the globe. Yet a few years later the entire system of hereditary monarchy—a previously essential institution—would be gone.

Let's review for a moment exactly what was being torn asunder: Not merely monarchy, but an entire social, cultural, and intellectual order. "Revolution" is overused and overhyped, but the era of the world wars ushered in a truly dramatic global transformation in just a few decades. Industrialization reached its zenith with automobiles and airplanes, science produced Einstein's theory of relativity and other dramatic breakthroughs, and we can also speak of the rise of professionalization, the beginnings of decolonization, suffrage for women (and with the flappers a new sort of liberation for women), the spread of mass communications, new movements in art and philosophy—the list goes on and on. All these new ways of thinking and new technologies worked together to destabilize Victorian norms and, amid the tumult of war, build the world we now inherit.

Astonished by "the sun of the old world . . . setting in a dying blaze of splendor," I kept reading and found myself fascinated by letters circulated between three first cousins: King George V of England, Kaiser Wilhelm II of Germany, and Tsar Nicholas II of Russia. These leaders grew up together, spending holidays with their grandparents—they adored (and were adored) by their grandmother, Queen Victoria—and penning intimate letters back and forth into adulthood. As the nineteenth century came to a close, the three imagined their great-grandchildren as the monarchs of Europe (and, indeed, the world) in our century. They were as blind as everyone to the changes that industrialization was bringing, and they were also blissfully ignorant of the growing dissatisfaction with hereditary monarchy. In one letter, a court observer from Russia writes:

> I must say there is life and merriment at the palace now. All the princely youth who have gathered here are letting their hair down. There is sunshine and laughter in all the corners of the palace, in all the corners of the park, as if a swarm of twittering birds had arrived.[3]

Sunshine and laughter—and meanwhile, Lenin was preaching revolution in the streets. Had they lived, the three cousins, "King, Kaiser, and Tsar," would not recognize the world by 1960, let alone today.

We're at the beginning of a similar epochal change in human history.[4] Scan the headlines every morning—through your Facebook and Twitter feeds—and you can feel history shifting under your feet. Every day I find more and more evidence that we are in the twilight of our own age, and that we can't quite grasp it, even if we can sense something is terribly amiss. This transformation transcends any one realm of life—it's all-encompassing, even if, as we've seen, it proceeds unevenly and paradoxically. Our twentieth-century institutions, which seem as foundational or ahistorical as hereditary monarchy, are on the cusp of collapse—or, if not outright collapse, of irrelevancy and anachronism.

Frequently these institutions have had a hand in their own demise. Then along came radical connectivity to hasten it by shifting immense amounts of power and influence toward everyday individuals—and to a few huge, Even Bigger platforms that dominate our digital world. What might have been a fifty- to one-hundred-year process has been compressed into a decade or even less. The End of Big replaces the elite, formal, highly capitalized, institutionally backed provider of goods or services with your neighbor the poet/journalist/lawyer/soldier/designer/(insert craft here). Soon, with 3-D printing, it won't just be media or intellectual property that anyone can create and disseminate; it will be anything—shoes, mobile phones, vehicles. And once anyone can create anything, brands and elite notions of excellence in any field really will be obsolete. It will all come down to relationships, to my neighbor (physical or digital) and my neighbor's work. If any of our institutions persist, it will be by virtue of sheer financial heft (although that didn't help monarchy), violence (also didn't work so well for the monarchy), or ingenuity. Let's hope for the last of these.

With the established order declining so quickly, people are getting anxious, depressed, and pessimistic. Over the last decade, doctors are writing more than twice as many prescriptions for anti-depressants, overwhelmingly for people without a diagnosed mental health condition.[5] The most popular stories of our age—Harry Potter, Batman, the Hunger Games, the proliferation of zombie films and TV shows—all tell of dark struggles between good and evil, the forces of good being tired or compromised in some way. A sense of "let's hunker down for the coming storm" pervades our social consciousness, captured well in the title of a poem by Charles Simic (one of my favorites): "Against Whatever It Is That's Encroaching." Barack Obama tapped into the ominous mood during his run for the presidency in 2008, articulating our hopes for a way out, a positive change. Guess what? The mood is even more ominous today.

Adding to our anxiety is the sense that we're powerless to disengage from the technology. Radical connectivity is addictive. We have trouble resisting the siren song appeal of the inbox, the social network, the incoming text messages. Whether at a conference or at home, we are glued to our devices, even when it puts lives at risk. According to several studies, texting while driving dramatically increases the risk of an accident, because reading or sending a text diverts the driver's eyes from the road for an average of 4.6 seconds—the same amount of time it would take to drive the length of a football field, blind, at 55 mph.[6] A 2009 experiment by *Car and Driver* magazine suggested that texting while driving was more dangerous than driving while drunk, based on measurements of stopping distance and response time.[7] And yet, despite the risks, the research, and the subsequent laws imposing heavy penalties, people continue to use phones while driving, in massive numbers.[8]

We face a monumental task helping our institutions recover from their failings so that they can serve their core purposes once again. Giving in to the anxiety and depression that accompany the decline of everything that makes our lives stable and safe seems a perfectly understandable response. But it's not the only response. In closing this book, I'd like to suggest that we can just as easily—and as realistically—think of the End of Big as an opportunity. No, the technology itself is not a miracle cure to everything that ails us, as some overly euphoric technologists think. But neither is it the insidious enemy enshrined in novels like *Frankenstein* and movies like *Terminator*. There is a middle ground. *We*, not the technology, can bring about the cure by assuming control of the technology, embracing where it is taking us while also having the collective determination and strength of mind to steer it where we want. Instead of lamenting the death of something ("The loss would be incalculable!"), we can learn to celebrate the creation of a radical new way of organizing the world, taking steps to align it with democratic values and with our need for social order.

Whenever I find myself feeling intimidated by the End of Big's chaotic face, I return to the stories of the founding fathers of the United States. All they had ever known was the monarchy of eighteenth-century Britain. There were rumblings of a revolution in France, but for these men it was unimaginably distant. They were for the most part wealthy white men, but many of them lived an agrarian existence, on busy farms. And yet they managed to imagine a way of doing things outside the existing institutions of their day. They worked hard to anticipate the ways it wouldn't work, to figure out the shortcomings of their proposed radical ways of doing things. And they also worked hard to bring the institutions of their imagination into existence in simple, practical ways.

My personal experience affirms for me that when we act deliberately, we can overcome technology's destructive effects to sustain dimensions of human experience that we care about. In her book *Alone Together,* Sherry Turkle has described technology's tendency to isolate individuals from one another even as it empowers them. She cites a study of more than 14,000 college students that found a dramatic drop in empathy over the last decade.[9] Turkle's own research suggests that in an effort to combat loneliness, we dive deeply into our devices—but not deeply into relationships. The optional nature of digital interaction makes relationships a one-sided choice and deprives us of the essential conflict that can form intimacy. As she reflects, "[connectivity] may assuage our deepest fears—of loneliness, loss, and death. . . . But connectivity also disrupts our attachments to things that have always sustained us—for example . . . face-to-face human connection."[10]

While I can attest to this disruption firsthand, I also have used technology to connect in powerful, human, off-line ways with people around me. In 2005, having recently been to Rome for the first time, I was reading a history of the Roman Republic and decided to write an excited post for my blog about the wonders of antiquity. A stranger named Dave Pentecost stumbled on to my blog and left an encouraging note. We ended up e-mailing back and forth and then met for lunch.

Dave seemed like a nice guy. At lunch, he invited me to join him on a trip to southern Mexico to visit archaeological sites related to Mayan antiquity, a longtime passion of his. He was leaving in two weeks. I went home, thought about his invitation, and decided to join him. Although we had only met once face-to-face, we had known each other online. An amazing adventure ensued and a life-long friendship was born. We even recorded a series of podcasts (which we titled "Junglecasts") exploring the wonders of Mayan antiquity, including interviews with archaeologists like the University of Texas professor Ed Barnhart. The podcasts soon took on a life of their own, receiving a small but nevertheless crazy number of downloads. This note came in via e-mail:

> Just letting you know these Jungle Podcasts are great and they are being heard as far away as Antarctica! Keep up the good work, cheers.
>
> Regards,
> Christopher R. Clarke
> Casey 2005 Expeditioner
> Wilkes Land, East Antarctica

My week among Mayan ruins and my subsequent friendship with Dave have added immeasurably to my life. From the Dean campaign to the Junglecasts and beyond, I have used the Internet deliberately—with a certain amount of clarity about my values and purposes—to enhance my life in rich and fantastic ways. I believe that we can do the same on a collective level, taking control of radical connectivity not to go backward but to recognize ourselves and our priorities once again in the new world of small it has ushered in.

As individuals and as a society, we need to acknowledge small as our future but simultaneously rediscover and embrace values such as limited government, the rule of law, due process, and individual freedoms of religion, speech, press, and assembly. While it may seem like a giant task, the first step is simple: start talking. A good way to staunch the decline of our existing institutional culture is by having a series of conversations with civic, political, and intellectual thought leaders who would in turn discuss the End of Big in their own ways with their audiences. Countless democratic efforts across the country would be emboldened if voices in the national media and Washington confirmed the discomfort most Americans feel with the state of our politics, our government, our commerce, our systems. Such confirmation would encourage Americans to become engaged in new and creative ways, on a grassroots level, feeding into a nationwide process of institutional renewal and reinvention. Talking would bleed quite naturally into *doing*.

I believe in human ingenuity and the great spirit of the American mind. I believe that the American people can craft and invent a new kind of culture, with the institutions and systems to match, that takes advantage of all of the technological and social advances of the last three centuries. Enshrining traditional democratic values in the institutions of the post-Big era won't be easy. Throughout the course of this book, we've seen a messy process taking shape whereby citizens are reassessing existing institutions, determining which to keep, which to modify, which to shelve. This process has bubbled up even as our government, locked in partisan gridlock, has moved toward a perilous combination of bankruptcy and irrelevance. We need to hasten this process and pursue it in a more deliberate, organized fashion.

In case calling for a new discourse about our institutions seems too vague, I'd like to help structure the conversation and subsequent efforts at reform by suggesting some ways of engaging that will help you and our country better inhabit the End of Big. First, *in revising institutions, focus on making them more amenable and responsive to individuals*. Every single citizen,

customer, client, employee, listener, reader, student, patient—
every person your organization touches—is powerful, almost
beyond measure. Treat them that way. Our institutions are
necessarily designed to subvert individuals, to bring order
through hierarchy, but it is not at all clear that this approach
makes sense anymore. We might as well embrace individuals
and make the most of what a more diffuse, nonhierarchical,
decentralized orientation has to offer. As Ori Brafman and Rod
Beckstrom wrote in their book *The Starfish and the Spider: The
Unstoppable Power of Leaderless Organizations,* "Decentralization
has been lying dormant for thousands of years. But the advent
of the Internet has unleashed this force, knocking down traditional businesses, altering entire industries, affecting how we
related to each other, and influencing world politics. The ab-
sences of structure, leadership, and formal organization, once
considered a weakness, has become a major asset."[11]

Second, demand serious, thoughtful, informed leadership. Most
politicians, educators, and business executives remain woefully unaware of the technology underlying radical connectivity. How many U.S. senators have ordered a book from Amazon?
Paid their bills online? Regularly watch YouTube? How many of
our university presidents know the difference between a
waiter and a (web) server?[12] Beyond just educating our leaders,
it's time to retire a lot of them. Thanks to the miracles of modern medical technology, the Baby Boomer generation is staying
active much longer than their forebears, retaining positions of
power across politics, government, education, and industry for
decades longer than is necessary or healthy.[13] That's not to say
we can navigate the future without the wisdom of our elders;
those who forget the past are doomed to repeat it. But leadership requires imagination and a willingness to bring about
disruption, and decades of power corrupt both the imagination

and the will. We need a new sense of public service—not just soulless meritocratic leaders, but leaders who can understand our challenges and inspire us to greater heights.

Third, we must *develop new processes and approaches to bring together the networked, individual power of radical connectivity and the direction-setting inspirational leadership necessary for change.* Barack Obama's 2008 presidential campaign offered an impressive example of how to harness the energy of empowered individuals while still providing clear leadership and direction. Obama expressed a vision and provided details around the vision. His leadership team provided tactical direction—how much money he needed to raise, what states he should focus on. Both the vision and the tactical direction were communicated transparently and openly online through measures such as YouTube video briefings with senior campaign staff members and regular e-mails to their list. Everyday Americans—more than 6 million of them—took the vision, leadership, and tactical needs of the Obama campaign and brought it into their local neighborhoods and online social networks. They hosted house parties, made phone calls, knocked on doors, made YouTube videos, wrote blog posts, and much more—and as a result, Obama won the White House.

Unfortunately, this impressive fusion of top-down leadership and distributed individual action across the network seemed to wilt once Obama actually came to occupy the White House. The reason for that is clear: The institutions of Washington, D.C.—namely the executive branch and the Democratic National Committee—are not nearly as flexible and malleable as political campaigns are. There are established ways of doing things, and they are broadly unwelcoming to the End of Big, as we have seen in the discussions of Big Parties (chapter 3) and Big Government (chapter 5). Still, in small ways the Obama team has made some headway, for instance by reinventing WhiteHouse.gov from a staid archive of press releases to a dynamic way of engaging with the American public.

As leaders become more sophisticated in their understanding of radical connectivity, our society can proceed to a fourth area of focus: *imagine in ever-finer detail what future institutions will look like*. Some existing institutions will simply go away; others might well be preserved in some form. As I outlined in chapter 2, the future of newspapers, our primary generators of accountability journalism, is bleak. They are going away, and we need to think about and imagine where accountability journalism will come from in the absence of journalism. Big Government, on the other hand, might be saved if we reorganize government and reimagine its process in the age of radical connectivity. The key to making our institutions relevant again is to look at them with new eyes, recognizing again that every person they touch—the people they serve, and the people who work within them—carries enormous personal power thanks to their digital life. How do we reimagine representative government with this insight in mind? How about big companies or big media? We must move institutions from the hub-and-spoke model—the "big will do this for you," the small, powerless individual—to new models that acknowledge and harness the individual's intense power and connectivity.

At the same time, we need to retain and reconstitute elements of the big institutions that we're losing: accountability journalism, national security, courts and justice, fiscal policy, safety regulations and quality control, to name a few. Resisting a radical, insular individualism, we must build institutions that encourage collaboration and accountability, locating such accountability in vast networks of small groups that share common culture and motives. Wikipedia is one model for this kind of distributed network with a shared common purpose and culture; open-source technical projects provide another model. Even the distributed chaos of the hacker collective Anonymous may provide a template for creating and shaping a new institutional force in the End of Big.

Beyond updating and salvaging our larger institutions, *we must strengthen and reimagine local community*—the fifth item on my suggested agenda. With the End of Big, our communities remain the fundamental building blocks of society. In fact, they have always reigned supreme; we've been misled these past few centuries by the delusions of the nation-state and global geopolitics. Clay Shirky has written a book about "cognitive surplus," the idea that people are using digital technology for creative acts rather than consumptive ones like watching television. The average American spends twenty-two to twenty-four hours a week watching television; you only need to devote a few of those hours to cultivating local community online or off-line to have an enormous impact. The notion of cognitive surplus underlies Wikipedia, the greatest collection of human knowledge ever assembled (even if it is incomplete!)—and it was also to some extent behind the 2008 Obama campaign's strategy. Take three or four hours a week and use the inherent power of your smartphone or laptop to contribute to a small-scale civic enterprise. The digital technology we carry around with us can be harnessed toward building new institutions (and reforming old ones) with surprising ease; we just have to watch a little less television.

Ultimately, small will save us—that is, if we commit ourselves to it. We can only hope to transform our current fossil fuel–based economy into a more sustainable system if we move collectively to small, sustainable, local energy sources. Food distribution today depends substantially on oil—flying in grapes from Chile, trucking oranges from Florida. We need to build more sustainable, local food production and distribution—a change that is already happening with the significant increase in farmers' markets and the rise of the local food movement. Our national economy will not resume its former glory if we aim policy making at Big Industry; large-scale manufacturing in the United States is to all intents and purposes dead and not returning. But the End of Big and the return to craft offers a potential pathway to sustainable economic growth. Moving to more local ways of approaching industry, commerce, agriculture, and government offers a future full of possibility—a bright future, one fully within our grasp.

A sixth and final way we can proceed in our new dialogue about institutions is to *take control of the Even Bigger platforms*— Facebook, Google, and Twitter—that today constitute our digital commons. These private companies have only the faintest realization of their civic role in providing a public space and see themselves as tools for individuals rather than as a digital town square. The history of newspapers is in part a history of print media companies slowly recognizing and coming to terms with their power and civic responsibility; we must hope—and demand—that the digital behemoths of our age manage a similar self-realization. And we must collectively consider what steps we should take to bring that self-realization about and, if all else fails, to enforce civic responsibility.

In her book *Consent of the Networked*, Rebecca MacKinnon documents a range of censorship and abuse—in some cases, in concert with governments—sustained by the large digital platforms. The Constitution assures us of freedom of speech in public areas like parks and roads, but it doesn't on Facebook, or Apple's App store, or any other corporate online space. MacKinnon elaborates:

The point of "consent of the governed" is that citizens trade absolute freedom to do anything and everything they please regardless of its impact on others, for a certain degree of security for themselves, their families, and their communities. This trade-off works reliably in the citizens' interest, however, only if government is held accountable—not only by a competitive political system and a strong, independent legal system, but also by constitutional constraints that guarantee respect and protection of citizens' most basic rights, which include free expression and assembly. . . . In the long run, if social networking services are going to be compatible with democracy, activism, and human rights, their approach to governance must evolve. Right now, for all their many differences, [digital platform companies like Facebook and Google] share a Hobbesian approach to governance in which people agree to relinquish a certain amount of freedom to a benevolent sovereign who in turn provides security and other services.[14]

MacKinnon is forging a new understanding of our digital life, building ideas like "consent of the networked," wherein a government's power and authority derives from the people it serves. Similarly, in consent of the networked, a technology platform's power is derived from its users—arguably much more so than a government's.

When you move to a country, you explicitly agree to its laws and the shape and form of its government. Once you're there, leaving it can entail a high transaction cost—even higher if the government doesn't want you to leave. When you join an online community, you're presented with the terms of service, an obtuse legal document frequently stretching on for pages. Facebook's terms of service are close to ten pages long. Many digital companies (like Yahoo!) change their terms of service without notification. These terms of service agreements read like anachronisms from an earlier age, using language that deserves greater discussion and review. When you click "Agree," you are consenting to participate in an online space under terms set by the company's management. You have no recourse should you find yourself accused of violating the terms of service, regardless of how vague those terms of service might be. That's not exactly "consent of the networked."

Doc Searls, a leading digital thinker, has an idea called vendor relationship management, or VRM. It's a long name for a simple idea: you should be able to decide your personal terms, and then all of the online platforms you use have to decide if they'll accept *your* terms of use. Doc explains what your terms might be:

- You may only collect data I permit you to collect.

- Any data you collect—for me, from me, or about me—is mine as well as yours, and will be made available to me in ways I specify (and here they are).

- You can combine my data with other data and share it, provided it is not PII (personally identifiable information).

- If we cease our relationship, you can keep my data but not associate any PII with that data.

- If we enter a paid relationship for services, you will spare me advertising and promotion for products or services other than yours. You will also not follow my behavior for the purposes of promotion or advertising. Nor will your affiliates or partners.

- You will put nothing on my computer or browser other than what we need for our own relationship. That includes cookies. (And here are the specific kinds of cookies I allow.)[15]

You could even demand that online platforms serve you the news and ideas of people you may not agree with to add some variety to your "information diet." Ideas like VRM are a crucial step forward in rethinking our relationship to our technology, as well as to the companies that manage the technology. VRM offers one possible means of attaining a "consent of the networked" and building new civic institutions.

While we need to acknowledge the role and power of these large digital platforms in our lives and communities, we must not overstate their power. In fact, it's relatively easy to leave an online service or community. Also, as Facebook grows and changes as a company (and faces pressure to generate more revenue) its stranglehold will dissipate, just as MySpace's did once quarterly revenue targets were introduced. Many theories exist about the Internet world's mass exodus from MySpace, but the fact remains that you don't need a substantial capital investment to build a compelling competitor. If I were to tell you today that a small team of smart, young nerds were building a Ford competitor, a car start-up that would take on America's best-selling automobile manufacturer, you would think I was crazy. But if I were to say that this smart, young team of nerds was building a Facebook competitor, you might be tempted to invest.

Yes, we've got these big technology platforms making decisions about our digital future, and, following MacKinnon's argument, we must build ways to hold these new emerging institutions accountable. But just as any smart, savvy motivated person with a laptop and Internet connection could pose a threat to national security, so, too, could that person pose a threat to established technology platforms like Facebook. Indeed, the pace of change in our technological world makes such platform challenges inevitable. The innovations we've come to know and love have happened so quickly, and destabilized so many of our core institutions, that the future remains astonishingly uncertain even for the companies that birthed our digital age. Large digital players like Google, iTunes, Amazon, and Facebook are critical players now, but they won't be for long. Within thirty-six months, a new giant will arise, as Twitter did, a few short years after Facebook, and Tumblr after Twitter. The technology simply offers no advantages to the established, to the institutional, to the victorious.

Technology platforms are also not as monolithic in their approaches as I may have suggested so far. In an interview with *Vanity Fair*, the 4chan.org founder Christopher Poole contrasted his approach with Facebook and Mark Zuckerberg:

> Mark's vision of the world is that you should be comfortable sharing as your real self on the Internet. . . . He thinks that anonymity represents a lack of authenticity, almost a cowardice. Though I like Mark a lot as a person, I disagree with that. . . . 4chan, a site that's anonymous and ephemeral, with wacky, Wild West–type stuff, has a lot to offer, and in Mark's perfect world, it probably wouldn't exist . . . He is a very firm believer that his is the right way for society to go.[16]

Despite a marked difference in worldview, both Facebook and 4chan revolve around individuals and the notion that individuals should have total freedom in anonymity, versus the individual having integrity in identity. Neither of these views accounts much for community, especially physical community. Other alternatives do—like Meetup.com, as detailed in chapter 3. My friend Max Novendstern has started the site OurCommonPlace, a social network designed around the people who live closest to you. It's a mini-Facebook, just for the floor of your apartment building or your neighborhood. It's not designed around casual updates, the way Facebook is, but around the mechanics and realities of physical communities. OurCommonPlace is one of many companies trying to figure out how to bring the radical connectivity of our technology into our physical communities with a civic bent, one that builds community and creates powerful, positive change.

Our Big Community—the civic life of our neighborhoods and our nation—has been in decline for the last fifty years, and here we have an opportunity to revive it with the End of Big. For the time being, we can't lose sight of the power of the Even Bigger technology companies that increasingly own the digital commons, the public space of the Internet. These large digital platforms of our time need to be held accountable, to understand their role as custodians of a civic space. They need to be about more than just algorithmically maximizing clicks in a manner that resembles addiction. The real challenge, however, is not just holding the emerging institutions accountable but designing new ones. Our new institutions must protect the values we hold dear, while also taking advantage of the wealth of networks, recognizing the power of the individual in our era of radical connectivity.

At first glance, the End of Big does seem dark, maybe even apocalyptic. Yet for a wide range of issues, from climate change to fighting corruption, the here and now of our local communities point to solutions to these challenges. We need the End of Big to bring us back to our communities, to our neighbors—that's how we'll remake the world and build a better future. Together, let's engineer structures that bring the hard-won values of the twentieth century together with the brilliant, game-changing technology of the twenty-first. It can be difficult to see the opportunity amid the chaos and fear that abounds, but make no mistake: The next decade will belong to those who can take the ground-up, grassroots energy unleashed by radical connectivity, marry it with effective, engaged leadership, and craft stable and responsive institutions. In other words, it will belong to those who gaze beyond the chaos of the End of Big, glimpsing one last big that stands unscathed: Big Opportunities.

AFTERWORD TO THE 2014 EDITION

When I awoke on April 15, 2013, to teach my Monday classes, the first thing I thought about was this book. Its release would take place in just eight days' time, and I found myself pondering once again its arguments and examples. Would readers agree with me? What kinds of questions would they ask online and at book events? With each passing day, it seemed that current events were bearing out, deepening, and on occasion also challenging elements of the story I had put down on paper some months earlier. This was exciting, but given what was happening around the world, somewhat disconcerting, too.

Arriving at the lecture hall at Harvard's Kennedy School, I arranged my notes and prepared to begin my presentation to the students. I looked around the room and remembered that one of my graduate students, Billy, was absent because he was running in the Boston Marathon. I went through my lecture and about eighty minutes later, having covered the necessary topics, I pulled up the marathon Web site on the big screen at the front of the classroom and typed in Billy's jersey number. His fellow students and I cheered him on, noticing that he was making good time and likely to get to the finish line around 3:00 P.M.

A couple of hours later, as I sat in my office, I noticed Twitter activity about a bombing at the Boston Marathon. At first, I didn't believe it. A photo was circulating of an incredible plume of fire on Boylston Street, just four miles away. The photo looked real, but I had seen this kind of thing doctored online before, and the people tweeting the photo were not people I knew but strangers being retweeted. A few minutes later, it became evident that something very bad had indeed happened. My thoughts rushed to Billy. It was 3:00 P.M. Doing some quick math, I figured that he was probably right at the finish line when the bombs went off. "Oh, crap," I said to myself. I logged on to Facebook and searched for his profile. He had just posted, via his mobile phone, a single word, "Okay."

Whew!

As the Boston bombing story unfolded over the following hours, days, and weeks, it illustrated the End of Big at every turn. It wasn't just the End of Big News—the dizzying stream of tweeted news and rumor that riveted me but also left me yearning for an authoritative source reporting on events. It was everything.

The bombers had been self-educated and radicalized through YouTube, blog posts, and other online sources, a terrifying reminder of the need to rethink national security in the face of challenges to Big Armies.

Everyday citizens were mobilizing technology to help in numerous ways, a development we might regard as the End of Big Philanthropy. Bostonians came out by the thousands to offer their homes to stranded marathoners using a Google spreadsheet. A friend of my wife's who went to high school with an injured victim started an online fund-raising campaign for him and his family, raising more than twenty-five thousand dollars overnight. Although Governor Deval Patrick and Mayor Tom Menino announced the creation of the One Fund Boston the day after the bombing, informal ad-hoc fund-raising for people affected by the bombing had already proliferated and gone viral online.

More than ever before, the marathon bombing affirmed how removed our leaders and institutions had become from the reality of our technology-saturated time. In an era when 130 million Americans have smartphones with the approximate power of a Cray Supercomputer, people were going to want to participate by helping to investigate the crime. And, in a development we can call the End of Big Policing (a subset of the End of Big Government), that's exactly what they did. Online communities like Reddit and 4Chan started crowdsourcing their own investigations of the Boston bombing, referred to as a "Racist Where's Wally" by one commentator. Two men were wrongly identified as suspects through the tangled web of online communities' attempts to help the investigation.

Our institutions of law enforcement were designed to bring integrity to the social body by following clear, rigorous processes aligned with democratic values. In this case, the FBI and Boston Police Department did in fact follow such processes, investigating the crime in a professional manner. But the desire of citizens to help enforce the law using radical connectivity isn't going away. Our institutions and leaders must figure out ways of channeling popular energy to extend the resources of law enforcement, just as it must do so to extend the resources of journalists and other professionals. Everyday citizens want to contribute to society in the moment and if we don't give them the chance to do so, they're going to find ways to "help" using today's technology that we might not like.

The FBI and the Boston Police provided a tip line via phone and an online form, but in this day and age, that is a primitive way of harnessing the power of crowds. Best practices for crowd-sourcing have become well established over the last decade: Define the problem, provide a lot of data, have a clear process to drive people through, and make sure you have a diverse network participating. Daniel Kreiss calls it "computational management," using technology, specifically web applications, to organize and drive volunteer energy in a specific direction and in a sophisticated way. Managing volunteers in this way is a far cry from merely having people e-mail their photos. Unfortunately, when the FBI and the Boston Police didn't provide a real process for an online community to help with the investigation, the online community did it anyway, with disastrous results.

It's true that the Boston Police Department was on Twitter, even using it to debunk misinformation that had gone viral. But here again, we find that critical opportunities were missed. Twitter could have—and should have—identified three or four "recommended" Twitter feeds to follow for good information on the Boston bombing: Perhaps the FBI, the Boston Police Department, the *Boston Globe*, and the mayor's office. While sitting in my office on Monday afternoon, four miles from the bombing, I was evacuated from the building due to rumors circulating on Twitter. Think of the economic disruption that ensues, not to mention the emotional trauma, when needless evacuations occur on a mass scale. Simply providing a slight nudge to direct people to the authorities can prevent the circulation and magnification of misinformation.

If aligning our institutions with emerging digital platforms like Twitter seems impossible at times, rest assured, a little bit of twenty-first-century-style Yankee ingenuity goes a long way. Keepr, an app born in the midst of the Boston bombing manhunt, allowed users to identify reliable sources that appeared to be tweeting from the scene. The app's creator, Hong Qu, was a visiting fellow at Harvard's Nieman Foundation for Journalism at the time of the bombing. His app used four factors to produce an automated credibility judgment: disclosure of location, preferably via verifiable geocoding, multiple source verification (i.e., determining if the tweets cited information from primary as well as other sources), original pictures or video, and accuracy over time. Keepr is a compelling example of how we might take some of journalism's core values and merge them with technology to bring greater integrity to the entire system.

The Thumb Drive Is Mightier Than the Sword

Looking beyond the Boston Marathon bombing, the transfer of power from traditional institutions to individuals continues at a brisk pace that our institutions cannot fathom. Perhaps the biggest End of Big story of 2013, among dozens, even hundreds I might cite, was the leaking of sensitive material by Edward Snowden as well as his evasion of the American authorities. In late January or possibly early February, Snowden started working with documentary filmmaker Laura Poitras to reveal what he had uncovered through his work as a government contractor with the National Security Agency (NSA). Poitras later introduced Snowden to Glenn Greenwald, a journalist with the British newspaper *The Guardian*, to break the story of Snowden's NSA leaks.

A figure like Snowden represents the End of Big personified. How is it that a young nerd, only on the job for three months, managed to bring the NSA—estimated to be the largest U.S. intelligence agency in terms of budget and staff—to its knees? We're talking about the most powerful intelligence entity in the history of the world, a central part of the national security apparatus of the United States, and it took Snowden three months, using a thumb drive and the Internet, to wreak havoc. We're only just beginning to understand the revelations of Snowden's massive leak; at a minimum, they proved the existence of several classified U.S. surveillance programs and their scope.

As of this writing, Snowden's leaks have had a substantial impact on international diplomacy and U.S. relations abroad. A recent story about the NSA eavesdropping on foreign leaders—even allies like German Chancellor Angela Merkel—is beginning to change U.S.-German relations. The Brazilian president Dilma Rousseff called off a visit to the United States over allegations that the NSA had eavesdropped on Brazilian officials, while the French government summoned the American ambassador over revelations that the NSA monitored phone calls in that country on a "massive scale."[1] Anger about NSA activities is also prompting a coalition of countries led by Brazil to oppose U.S. control (via the Department of Commerce) over the underlying protocols on which the Internet is built. Rousseff announced plans to create an undersea fiber-optic cable that would funnel Internet traffic between South America and

Europe, bypassing the United States entirely. The Brazilian legislature is considering an amendment that would force Google, Microsoft, and other Web companies to store data for Brazilian users on Brazil-based servers. The Brazilian postal service is even developing a "more secure" alternative to e-mail that would be entirely homegrown and safely encrypted from any spying eyes.

As in the case of Big Political Parties and Big Media, we also have the specter in the Snowden case of a profoundly failing institution going horribly awry. The United States has a long tradition of civilian oversight of the military and of the separation of powers, in large measure so that the different branches of government can hold one another accountable. Thanks to Snowden, we now understand how the NSA has wielded excessive power, watching the online communications of everyday Americans, without any accountability. In March of 2013, just a few weeks before Snowden's leaks, Senator Ron Wyden asked Director of National Intelligence James Clapper if the NSA was spying on millions of Americans. This was the U.S. Senate doing its job of holding the executive branch and the military-intelligence agencies of the United States accountable to the American people. What did Director Clapper do? He lied. When asked point-blank by Senator Wyden if programs like PRISM existed, Clapper responded, "No."

Following Snowden's leaks, Andrea Mitchell interviewed Clapper on MSNBC and asked him about how he had answered Senator Wyden. Clapper told Mitchell, "I responded in what I thought was the most truthful, or least untruthful, manner by saying 'no.'"[2] The "least untruthful" answer was an astonishing evasion of the accountability the Constitution demands. It is exceptionally dangerous for any country to have an unelected and unmonitored group inside the government with the kind of surveillance powers possessed by the NSA. As George Orwell reminds us, power corrupts, and absolute power corrupts absolutely.[3] Yet the End of Big confronts us with a paradox: Even as the NSA has grown more powerful, a single individual can still instantly disrupt it with the help of technology. To Orwell's saying we might add another, equally relevant one: The thumb drive is mightier than the sword.

The Death of Privacy

The increasing role of government surveillance on social networks will not surprise anyone with experience in the digital realm over the last decade. As I outlined in chapter 9, a handful of Internet companies control most of our experience online: Amazon, Apple, eBay, Facebook, Google, Skype (owned by Microsoft), and Twitter. These seven companies—and many others built atop their technology platforms—have to a large extent built their business models around monitoring consumer behavior online. Any expectation we might have of privacy has died a slow death as we've willingly traded our privacy for the utility of Google Search and the fun of perusing friends' Facebook updates.

If we assume privacy is lost irrevocably, then perhaps it's time to ask the question: Is privacy an essential part of democracy? I would argue that it still very much is. Consider voting. Every year on the Tuesday after the first Monday in November, Americans arrive at their polling stations and line up. If they're traveling or living abroad, they post their vote to their home precincts to be counted. Wherever Americans are, however they vote, one thing holds true: Their ballots are held in secret.

Secret ballots didn't always exist in the United States.

Americans used to vote publicly, either aloud or with a physical movement such as the raising of a hand or the moving to one side of a room. Such public ceremony left voters open to intimidation or corruption, affecting the integrity of our elections. Inspired by our Australian friends, our leaders instituted the secret ballot, among other rules. By 1892, voting was private, beyond surveillance, allowing for citizens to exercise their constitutional rights to vote free of undue manipulation. The secret ballot was never an end in itself but rather a way to protect this right. Reformers understood how important it was for individuals to be able to be able to make their own decisions freely and safely, to vote their consciences irrespective of what their employers, elected officials, or others in positions of power thought.

Sadly, our big-data landscape hasn't caught up with 1892. The legal definition of "personally identifiable information" used by the U.S. Office of Management and Budget does not include several categories, including school or place of residence, which could be combined with other data to identify you. Rapidly evolving technology has made it possible for big entities of all kinds—campaigns, corporations, even governments—to exert influence on targeted individuals. The threat posed by algorithmic personalization, pervasive monitoring, and huge volumes of data is subtle. The basic rights we hold dear—free speech, peaceful assembly, due process—do not vanish when we go online, but the digital era is chipping away at their foundations by enabling a variety of actors to exploit our attributes to shape our choices for their private gain.

Ethan Roeder, the former data director of Obama for America, has claimed that "New technologies and an abundance of data may rattle the senses, but they are also bringing a fresh appreciation of the value of the individual to American politics."

Roeder means well, but we must ask an important question: value for whom? Current privacy standards aren't adequate to protect citizens in a democracy. We are overdue for a serious examination of long-term solutions that protect our right to be the true author of our lives. Our old institutions aren't up to the task, but the big digital platforms of our digital age aren't going to do it either. We—the citizens—will have to do that ourselves.

End of Big Banks?

The stories of the Boston bombing and Edward Snowden raise a host of issues about the End of Big News, Big Armies, and Big Government. But as I've traveled around the world to talk about the book, the greatest skepticism I've encountered to the idea of the End of Big has had to do with Big Business, and most notably entrenched financial interests. People observe the huge profits reaped by large corporations and doubt whether economic power has shifted to individuals, as I contend. I invariably get asked a question that goes something like this: "How can you talk about the End of Big when a handful of banks are bigger than ever, and when their reckless actions have precipitated the worst economic crisis since the Great Depression?"

I think it's clear that some of the economic or financial trends I outlined in this book have grown and intensified. The best examples surround the rise of alternative currencies like Bitcoin and the expansion of the "sharing" (i.e., non-money) economy. As of October 2013, the entire Bitcoin monetary supply is equivalent to approximately $1.5 billion U.S. dollars. There are Bitcoin hedge funds and currency exchanges, and at least two major U.S. investment groups have made substantial investments in Bitcoin. There is even a Bitcoin exchange traded fund (not yet approved by the SEC but in progress) backed by the Winklevoss twins of Facebook fame. For Bitcoin, such momentum is as much a liability as it is an asset. Bitcoin is beginning to attract significant regulatory review, with a federal judge in Texas ruling that Bitcoin is a legitimate currency.

Another trend I identified as a threat to Big Business was the rise of 3D printing and on-demand fabrication. Shortly after the initial publication of this book, Defense Distributed, an organization in Texas, published online the blueprints to print magazines for the AR-15 and the AK-47 assault rifles as well as for the Liberator, the world's first fully 3D-printable gun. In 2014, a range of important patents for 3D printing expire, and when they do, we're likely to see a proliferation of 3D printers for on-demand fabrication. Already MakerBot, the company profiled in chapter 8, has released a 3D scanner, which allows for some pretty intriguing possibilities. Do you like your friend's shoes? Borrow them, stick them in the scanner, and then print a copy. Current technology is rough and has its drawbacks, but I still contend that the pace and direction of 3D printing promises to revolutionize manufacturing.

Alternative currencies, the sharing economy, and 3D printing in this book are all nascent trends. In their current form, at their current size, they do not threaten Big Business. But all three of these trends challenge fundamental assumptions about our economy, dislodging many economic functions from the sole grasp of large organizations. In the case of banking, if I am able to live at a reasonably high quality of life using current and emerging technologies, I can essentially bypass the existing, entrenched financial system (or at least engage with it minimally). As the existing "Big Banks" system continues to abuse its power and resist accountability, alternative systems look more and more compelling to those not in on the giant profit margins generated by investment banks. Yes, the Big Banks are still very Big, and very powerful. But they are simultaneously more fragile than we think. Just like monarchy at the dawn of the twentieth century, their days are numbered.

Netflix Gets Big

As Big Business in general struggles to make sense of changes in the economy, the entertainment industry is reeling from titanic changes. In July 2013, Netflix received fourteen nominations for Primetime Emmy Awards, nine of them for the series *House of Cards,* only available on the online service. Netflix also recently surpassed cable giant HBO in number of subscribers. Or look at Kickstarter, profiled in chapter 5, which has since crowdfunded $5 million for a Veronica Mars movie that would not otherwise get made. This isn't just a single big outlier: It turns out that more than half the successfully funded projects on Kickstarter are for music, film, and art. Whether it is Netflix or Kickstarter, the traditional layer of "Big," organized intermediaries that help filter, fund, and cultivate talent to create big hits is under siege.

Unfortunately, much of the revenue lost by traditional players in the entertainment-industrial complex is not going to artists. It's going to different, sometimes even Bigger players: the digital platforms like Amazon, Apple, and Netflix. And let's not forget that the Internet, for all the direct access it allows between producers and consumers of cultural products, also brings with it a different, instant kind of censorship and control. Remember when "cartoon boobs" led to Apple removing an illustrated graphic novel version of James Joyce's *Ulysses* from its iTunes store? We need existing entertainment institutions to step in and experiment with new vehicles and models for funding, creating a new kind of intermediary—perhaps starting with the curated pages on Kickstarter. Digital platforms that seek to replace traditional cultural players (like Amazon's self-publishing) must also build better ways of encouraging and supporting artists. Someone needs to focus on the cultivation, care, and feeding of talent.

Still, we can't rely only on the institutions, traditional or new, to put more intentionality into the future we're building for artists. Artists have always struggled to make money and find an audience. But it's not sustainable to have an "industry of one" for every artist. We—both artists and fans—are going to have to figure this out ourselves.

Shutdown

In the fall of 2013, the United States Federal Government shut down for sixteen days. Eight hundred thousand federal employees were furloughed; 1.3 million federal employees required to show up for work were not paid. Standard & Poor's estimated that the shutdown took $24 billion out of the U.S. economy. At the root of the shutdown was a dispute between President Barack Obama and House Republican leaders. But the nature of the dispute was murky; House Republican leaders did not have a clear set of demands, but rather a general opposition to the president on every front. Unable to pass a continuing resolution (which is a sort of budgeting vehicle, a way to agree on how much money to spend on what), the shutdown dragged on.

Another deadline loomed: In order to pay its bills and avoid defaulting on its credit obligations, the government needed to raise a legal limit or "ceiling" on how much debt it could carry. But House Republicans balked on this issue, too. Finally, at the eleventh hour, a minority of Republicans and a majority of Democrats managed to break the impasse, coming to an agreement to raise the debt ceiling for ninety days and restore the federal government to regular operation. A dramatic crisis had been averted, but the deal ensured that the entire drama would be repeated just after the New Year.

At the heart of the affair was a young, first-term senator from Texas, Ted Cruz. Representing the Tea Party wing of the Republican Party, Cruz had helped produce the crisis by urging his former colleagues in the House to resist any compromise with the president. His actions had incurred the wrath of his Senate Republican colleagues, with Senator John McCain calling Cruz's position "shameful." Polls showed that the vast majority of Americans were critical of the government shutdown, and that they blamed the Republican Party. Some commentators and political prognosticators were predicting that the shutdown put into jeopardy the Republican control of the House, a previously unthinkable proposition. None of this nor the damaging of American prestige on the world stage stopped Cruz from taking a "victory lap" around Texas and Iowa to celebrate what he regarded as a giant success. For Cruz, whose visit to Iowa fanned rumors of a possible presidential run, the shutdown was a badge of honor and a measure of victory. Obstructionism and nihilism ruled the day.

The Tea Party's role in shutting down the government is a natural outgrowth of the trends I covered in chapter 3 on Big Political Parties. Ted Cruz and his Tea Party colleagues do not feel accountable to the traditions of the U.S. Senate or even those of the Republican Party. Thanks to radical connectivity and the nature of our technology, they have their own, independent power base. They raise money online, outside of the traditional Republican Party, and organize separate meetings and events. They even have their own media—a "filter bubble" of news mixed with opinion. Radical connectivity empowers insurgents at the expense of the establishment. In politics, a range of institutional changes (including the repeal of the earmark, which deprived House Republican leaders of any leverage in their negotiations with rogue members of Congress) has only exacerbated this phenomenon, as has the weakness of existing institutions, which makes alternatives organized via technology more attractive. The future only promises more of this disruptive, dangerous approach to governance. Get ready for a wild ride.

Big Web sites

While the Tea Party was bringing our country to the brink of economic ruin, the Democrats seemed unable to implement President Obama's crowning achievement, health-care reform. On October 1, 2013, the White House launched HealthCare.gov to much fanfare. Almost immediately the site crashed. A month after being launched, the Web site's widespread problems forced even the President's supporters to acknowledge it as, at best, a "fiasco." Yet these "glitches" should come as no surprise. In the age of radical connectivity, "Big" is a liability. The Big Institutions—huge federal bureaucracies, large technology companies, and a massive, unwieldly government procurement process—spent hundreds of millions of dollars to produce an administrative mechanism for health care, and they failed miserably. Smaller and nimbler technologies and processes could have prevented these problems, which threatened the success of the president's entire, legacy-defining legislation.

Meanwhile, another large, important Web site created untold stress for tens of thousands of people. More than five hundred leading universities use the same application, called the Common College Application, to vet prospective students. But while the application itself has been around more than thirty years, the recent launch of a new online system has been plagued by issues, leading one college admissions expert to refer to it as "application Armageddon."[4] As someone who has built Web sites for close to twenty years, I find it surreal to see a national conversation about a Web site failure, let alone two at the same time. A recent study, released last week by *Computerworld*, revealed that 96 percent of IT projects with budgets of greater than $10 million fail. Big just doesn't scale.

An oft-repeated mantra in the open-source world of technical development is "release early and often." Start small, make sure your assumptions work, and build iteratively towards scale. Most technical leaders I know would never allow a massive public launch of a highly touted program without smaller phased rollouts. The attitude and approach to both HealthCare .gov and the Common College Application make it clear that a growing gap exists between traditional institutions like the federal government and our largest universities and the technology most Americans use to navigate their lives. If our leaders cannot close that gap, our faith and trust in institutions will continue to fall, and we will keep disengaging from them to seek alternatives that work better.

The failure of HealthCare.gov and the Common College Application support one of my conclusions in chapter 9: that we must demand that our leaders have technical literacy, if not technical fluency. Most politicians, educators, and business executives remain woefully unaware of what kinds of questions to ask to hold contractors and programmers accountable. In the past, all political problems were people problems—bad policy decisions, personnel scandals, wars, corruption, and the like. Now software problems have become political problems that endanger the signature policy achievement of a presidential administration. And yet it appears that no one at high levels in the White House was monitoring the development of HealthCare.gov with an eye on the political impact of a failure.

Some Blessed Hope

Radical connectivity is pushing power to individuals and away from established institutions at an accelerating pace. This paves the way for a breathtaking next couple of decades: exciting for the promise and opportunity they hold, but terrifying for the chaos and upheaval that are likely. Overall, I found the writing of this book a hopeful experience. I began writing with a dark view of what the future might hold. The trends I saw taking shape were powerful and dangerous, and our leaders and institutions seemed oblivious. As I spoke to people around the world, and followed various threads to their conclusion, a different picture gradually took shape. Yes, our world is not prepared for the change technology is bringing. But some individuals *are* prepared, and they're making a difference. Wherever in the world I've gone to talk about this book, I've found an inspired few who are taking the power technology has given them and creating something new, something better, something different.

At the end of the nineteenth century and the start of the twentieth century, the great writer Thomas Hardy could tell that the big institutions of his era were in decline and facing a tumultuous future. And yet he was able to see some evidence of hope that something better was out there, a future even beyond his wildest, most ecstatic expectations. On December 29, 1900, he published a poem about a small bird singing in a desolate landscape in the middle of winter, titled "The Darkling Thrush." The final stanza describes Hardy's reaction to the bird's warm song:

> So little cause for carolings
>   Of such ecstatic sound
> Was written on terrestrial things
>   Afar or nigh around,
> That I could think there trembled through
>   His happy good-night air
> Some blessed Hope, whereof he knew
>   And I was unaware.

NOTES

For more details, visit endofbig.com/endnotes

1. Burn It All Down

1. Philip Levine, "M. Degas Teaches Art and Science at Durfee Intermediate School, Detroit 1942," *What Work Is: Poems* (New York; Alfred A. Knopf, 1991), 38.

2. A little reminder as to how breathtaking our technological capacities have become: In 1949, my grandfather took a business trip to North Africa. After a month, he wrote my grandmother a letter asking her to meet him in Paris on a specific date. She left their home in San Francisco and arrived on that date, excited to see him. Yet he never showed. He didn't come the next day, either. Or the day after that. She waited

and waited. She couldn't call him, because there were no mobile phones in those days, and even landlines in North Africa were scarce. A week later, he finally showed up. I once asked her if she was anxious; I would have assumed something terrible had happened. "No," she said, "in those days you would just wait." Such waiting is inconceivable now, thanks to radical connectivity.

3. http://www.workplaceethicsadvice.com/2011/10/flash-mobs-threaten -retail-industry-retailers-are-facing-a-new-threat-this-holiday-season -swarms-of-teenagers-and-young-adu.html

4. A number of thinkers and writers have begun to explore the implications of our technology, perhaps most notably Sherry Turkle, Jaron Lanier, and Clay Shirky, all of whom have different points of view on the subject.

5. Jaron Lanier, "The Hazards of Nerd Supremacy: The Case of WikiLeaks," *The Atlantic*, 20 Dec. 2010.

6. http://www.mcsweeneys.net/articles/in-which-i-fix-my-girlfriends -grandparents-wifi-and-am-hailed-as-a-conquering-hero

7. With apologies to Benjamin Nugent, author of *American Nerd: The Story of My People*.

8. In 1965, the Intel cofounder Gordon Moore suggested that every eighteen months, computer chips would become twice as fast, half as expensive, and half as big. It turned out to take more time: Every twenty-four months computer chips shrink to half their size while getting twice as fast and twice as cheap.

9. http://www.dougengelbart.org/history/engelbart.html

10. Ibid.

11. In May 1970, a group of students at the University of Illinois organized a day of action to protest the construction on campus of a supercomputer called the ILLIAC IV, primarily because it was funded by the Defense Department. They called their protest Smash ILLIAC IV and included a cartoon of the mainframe computer with screens tracking things like a "kill-die factor" and a gaping mouth labeled "Feed $$$$$$ here!"

12. Stewart Brand is a particularly interesting figure because he bridged these two branches of nerd culture. He was the camera operator at Engelbart's "Mother of All Demos," but he was also one of the Merry

Pranksters running around on Ken Kesey's bus. The quotation is taken from his essay "We Owe It All to the Hippies," *Time*, 1 Mar. 1995.

13. http://www.digibarn.com/collections/newsletters/peoples-computer /peoples-1972-oct/index.html

14. http://www.atariarchives.org/deli/homebrew_and_how_the_apple .php

15. http://www.digibarn.com/collections/newsletters/homebrew/V2_01 /index.html

16. http://www.gadgetspage.com/comps-peripheral/apple-i-computer-ad .html

17. John Markoff, *What the Dormouse Said: How the Sixties Counterculture Shaped the Personal Computer Industry* (New York: Penguin, 1996).

18. http://pdgmag.com/2012/02/02/steve-jobs-lee-clow-and-ridley-scott -the-three-geniuses-who-made-1984-less-like-1984/

19. http://www.youtube.com/watch?v=OYecfV3ubP8

20. Adelia Cellini, "The Story Behind Apple's '1984' TV Commercial: Big Brother at 20." *MacWorld* 21, no. 1 (2004): 18.

21. http://partners.nytimes.com/library/tech/99/12/biztech/articles /122099outlook-bobb.html?Partner=Snap

22. Noam Chomsky and others have argued that the United States has a long-standing tradition of using the Pentagon as a cover for massive government investment in strategically important industries. Calling something a defense expenditure makes it much more less vulnerable to political attack.

23. Mitch Kapor may not be a household name—but he should be. Many of you may remember Mitch alongside Steve Jobs and Bill Gates as one of the legends of the personal computer. He founded Lotus Development Corporation, an early software company, and pioneered applications for spreadsheets and graphics. He has gone on to have a leading role in a number of other important organizations, including as a cofounder of the Electronic Frontier Foundation.

24. https://projects.eff.org/~barlow/Declaration-Final.html

25. http://www.vanityfair.com/culture/features/2008/07/internet200807

26. Having two presidential candidates for 310 million people is wildly

undemocratic. At this point, you can't have a democratic (small *d*) conversation about the presidency, but you can help elect him (and it is generally a him) in a more democratic way than before.

27. Doug Casey, "End of the Nation-State," *LewRockwell.com*, 4 Jun 2012. http://lewrockwell.com/casey/casey123.html.

28. Liza Hopkins, "Citzenship and Global Broadcasting: Constructing National, Transnational and Post-national Identities," *Continuum: Journal of Media and Cultural Studies* 23, no. 1 (2009).

29. http://money.cnn.com/2011/07/13/news/companies/vivek_kundra _leadership.fortune/index.htm

2. Big News

1. Jim Harrison, letter 27, *Letters to Yesenin* (Port Townsend, WA: Copper Canyon Press, 2007), 53.

2. http://blog.socialflow.com/post/5246404319/breaking-bin-laden-visu-alizing-the-power-of-a-single

3. http://twitter.com/#!/ReallyVirtual/status/64780730286358528.

4. http://www.marketingcharts.com/print/just-two-in-five-americans -read-newspaper-daily-11646/; see also http://pewinternet.org/Reports /2011/Local-news/Part-1/Age.aspx

5. http://scripting.com/stories/2009/05/15/sourcesGoDirect.html

6. Alex S. Jones, *Losing the News* (New York: Oxford University Press, 2009).

7. http://www.cjr.org/essay/confidence_game.php?page=all

8. http://www.people-press.org/2012/09/27/in-changing-news-landscape -even-television-is-vulnerable/

9. http://sethgodin.typepad.com/seths_blog/2006/10/five_common_cli .html

10. http://online.wsj.com/article/SB1000142405270230444780457641483286 9865182.html

11. http://www.sifry.com/alerts/archives/000245.html

12. http://www.invesp.com/blog/business/how-big-is-blogsphere.html

13. http://www.huffingtonpost.com/2010/01/15/medianews-group-files -for_n_425542.html

14. http://www.utsandiego.com/news/2009/feb/23/newspapers -bankruptcy-022309/ (AP)

15. http://www.good.is/post/sun-times-media-group-goes-bankrupt/

16. http://www.nytimes.com/2011/12/13/business/lee-enterprises-files -for-bankruptcy-protection.html

17. http://scripting.com/stories/2009/05/15/sourcesGoDirect.html

18. http://a.nicco.org/zVWo0q

19. In January of 2003 former Colorado Senator Gary Hart briefly blogged at Gary Hart News.com as part of a short-lived consideration for another presidential campaign. Source: http//techpresident.com/timeline

20. http://www.alternet.org/module/printversion/29833

21. http://www.journalism.org/analysis_report/covering_great_recession

22. http://www.niemanwatchdog.org/index.cfm?fuseaction=background .view&backgroundid=00409

23. http://www.cjr.org/feature/money_talks_marchapril2012.php ?page=all

24. http://www.nakedcapitalism.com/2007/01/when-did-housing-lending -standards.html

25. http://www.nytimes.com/2007/01/28/realestate/28mort.html?_r=1 &scp=4&sq=lending&st=nyt

26. http://stateofthemedia.org/2011/cable-essay/data-page-2/

27. http://www.stevenberlinjohnson.com/2009/03/the-following-is-a -speech-i-gave-yesterday-at-the-south-by-southwest-interactive-fes tival-in-austiniif-you-happened-to-being.html

28. Alex S. Jones, *Losing the News* (New York: Oxford University Press, 2009), 4.

29. http://digitalhks.pbworks.com/w/page/50872581/Media%3A%20Texas %20Tribune

30. http://www.niemanlab.org/2009/06/four-crowdsourcing-lessons-from -the-guardians-spectacular-expenses-scandal-experiment/

31. http://mashable.com/2010/11/24/investigative-journalism-social-web/

32. Amanda Michel, "Get Off the Bus: The Future of Pro-Am Journalism," *Columbia Journalism Review*, Mar.-Apr. 2009: http://www.cjr.org/feature /get_off_the_bus.php.

33. Michel, "Get Off the Bus."

34. http://www.propublica.org/special/reportingnetwork-signup

35. http://www.kk.org/thetechnium/archives/2008/03/1000_true_fans.php

36. http://www.cjr.org/feature/the_josh_marshall_plan.php

37. http://www.huffingtonpost.com/jeff-cohen/josh-marshall-on-the
-grow_b_131571.html

38. http://www.huffingtonpost.com/jeff-cohen/josh-marshall-on-the
-grow_b_131571.html

39. http://www.niemanlab.org/encyclo/talking-points-memo/

40. http://en.wikipedia.org/wiki/Wikipedia:Five_pillars

41. http://blog.oregonlive.com/mapesonpolitics/2010/04/kitzhaber_oppo
nent_of_gorge_ca.html

42. http://storify.com/acarvin/rep-gifford

43. http://storify.com/jcstearns/tracking-journalist-arrests-during-the
-occupy-prot

3. Big Political Parties

1. Jack Gilbert, "Failing and Flying," *Refusing Heaven* (New York: Alfred A.
Knopf, 2005), 18.

2. http://www.nytimes.com/2000/02/13/weekinreview/the-nation
-courting-web-head-cash.html

3. Candice Nelson, *Grant Park: The Democratization of Presidential Elections,
1968–2008*, Washington, D.C., Brookings Institute Press, 2011. See also
Peter Goldman and Tony Fuller, *The Quest for the Presidency 1984* (New
York: Bantam Books, 1985), 142-144; Jack W. Germond and Jules Wit-
cover, *Wake Us When It's over: Presidential Politics of 1984* (New York:
Macmillan, 1985), 163.

4. http://news.newamericamedia.org/news/view_article.html?article_id
=8f096495b5b5ff31ea2e41abfa7d00a3

5. http://www.nytimes.com/2012/07/13/opinion/brooks-why-our-elites
-stink.html?_r=2&ref=davidbrooks

6. http://www.cjr.org/campaign_desk/how_to_understand_the_invisibl
.php?page=all

7. http://www.pbs.org/newshour/gergen/july-dec99/drew_7-23.html

8. http://www.pbs.org/newshour/convention96/retro/beschloss_history.html

9. Ted Sorensen, *Kennedy: The Classic Biography* (New York: HarperCollins, 2009), 128.

10. Qtd. in Christopher Matthews, *Hardball: How Politics is Played—Told by One Who Knows the Game* (New York: Simon and Schuster, 1999), 155.

11. John Aldrich, "The Invisible Primary and Its Effects on Democratic Choice." *PS: Political Science and Politics* 42, no. 1 (2009): 33–38.

12. http://www.pbs.org/newshour/gergen/july-dec99/drew_7-23.html

13. http://www.npr.org/blogs/money/2012/03/30/149648666/senator-by-day-telemarketer-by-night

14. http://www.nybooks.com/articles/archives/2012/mar/22/our-corrupt-politics-its-not-all-money/

15. http://www.mendeley.com/research/theory-political-parties-2/; see also http://a.nicco.org/L3wbUY

16. In John Tedesco and Andrew Paul Williams, *The Internet Election: Perspectives on the Web in Campaign 2004* (Lanham: Rowman and Littlefield, 2006).

17. Qtd. in James Barnes, "For Now, the Joke's on the Establishment," *National Journal* 30 Aug. 2003.

18. McCain, although formerly a maverick candidate, by the 2008 presidential campaign had come to wear the establishment mantle.

19. http://www.huffingtonpost.com/2012/07/21/2012-ads_n_1691942.html

20. http://journals.cambridge.org/action/display Abstract?from page-on line&aid-1818260

21. http://www.brookings.edu/papers/2010/04_polarization_galston.aspx

22. http://nationaljournal.com/magazine/divided-we-stand-20120223

23. http://a.nicco.org/L3zPOA

24. http://a.nicco.org/L3zMlY

25. http://a.nicco.org/Jb57qY r

26. http://www.rollingstone.com/politics/news/michele-bachmanns-holy-war-20110622#ixzz1vE8s5oY3

27. Qtd. in http://www.newyorker.com/reporting/2011/08/15/110815fa_fact_lizza#ixzz1vE985NIX

28. http://www.youtube.com/watch?v=7GSmDsAET7I

29. http://thecable.foreignpolicy.com/posts/2012/05/11/hagel_reagan_wouldn_t_identify_with_today_s_gop

30. https://secure.actblue.com/about

31. http://www.cnn.com/2011/11/07/tech/web/meetup-2012-campaign-sifry/index.html

32. http://www.wired.com/wired/archive/12.01/dean.html

33. http://www.nytimes.com/2009/09/06/jobs/06boss.html?_r=1

34. Ibid.

35. http://en.wikipedia.org/wiki/Americans_Elect#Ballot_status

36. http://www.americanselect.org/

37. http://www.nytimes.com/2012/05/06/sunday-review/direct-democracy-2-0.html

38. http://transparency.globalvoicesonline.org/

39. http://www.newyorker.com/reporting/2011/04/04/110404fa_fact_ioffe?currentPage=all

40. http://www.bbc.co.uk/news/world-europe-16057045

41. http://online.wsj.com/article/SB10001424052970203986604577257321601811092.html

42. http://techpresident.com/news/wegov/22198/culture-hacking-how-one-project-changing-transparency-chile

43. http://www.nytimes.com/2012/03/07/business/web-sites-shine-light-on-petty-bribery-worldwide.html?pagewanted=all

44. http://sunlightfoundation.com/blog/2009/05/26/brandeis-and-the-history-of-transparency/

45. http://influenceexplorer.com/

46. http://www.nytimes.com/2010/01/24/magazine/24fob-wwln-t.html

4. Big Fun

1. Elizabeth Bishop, "I Am in Need of Music," The Complete Poems 1927–1979 (New York: Farrar, Straus and Giroux, 1984), 214.

2. http://www.forbes.com/sites/michaelhumphrey/2011/05/31/shaycarls-epic-journey-to-youtube-stardom/

3. http://www.youtube.com/t/press_statistics

4. http://www.youtube.com/t/press_statistics

5. Qtd. in Scott Kirsner, *Inventing the Movies: Hollywood's Epic Battle Between Innovation and the Status Quo, from Thomas Edison to Steve Jobs* (Boston: CinemaTech Books), 199.

6. http://www.nytimes.com/2012/04/27/nyregion/at-92-movie-bootlegger-is-soldiers-hero.html?pagewanted=all

7. http://www.sfgate.com/news/article/Assessing-Napster-10-years-later-3229454.php#page-2

8. Qtd. in Robert Levine, *Free Ride: How Digital Parasites Are Destroying the Culture Business, and How the Culture Business Can Fight Back* (New York: Doubleday, 2011) 38.

9. According to Forrester Research via CNN (http://money.cnn.com/2010/02/02/news/companies/napster_music_industry/)

10. http://www.tc.umn.edu/~jwaldfog/pdfs/American_Pie_Waldfogel.pdf; http://papers.ssrn.com/sol3/papers.cfm?abstract_id=1944001

11. http://www.tc.umn.edu/~jwaldfog/pdfs/American_Pie_Waldfogel.pdf

12. Qtd. in Levine, *Free Ride*, 40.

13. Qtd. in Josh Tyrangiel, "Radiohead Says: Pay What You Want," *Time* 1 Oct. 2007. Retrieved 16 Oct. 2007.

14. Amazon now has its own self-publishing, print-on-demand service that competes with LuLu.

15. The book is available for purchase at http://www.bulldozingtheway.com/.

16. https://buy.louisck.net/news/a-statement-from-louis-c-k

17. https://buy.louisck.net/purchase/live-at-the-beacon-theater

18. http://www.gq.com/entertainment/tv/blogs/the-stream/2012/03/aziz-ansari-dangerously-delicious-standup-online.html

19. http://bigthink.com/ideas/42326

20. Kickstarter statistics are constantly updated at http://www.kickstarter.com/help/stats.

21. http://www.kickstarter.com/blog/10000-successful-projects

22. http://mediadecoder.blogs.nytimes.com/2012/01/30/at-sundance-kickstarter-resembled-a-movie-studio-but-without-the-egos/

23. http://www.nytimes.com/2008/06/22/magazine/22madmen-t.html

24. Levine, *Free Ride*, 141.

25. http://www.slate.com/articles/business/moneybox/2012/02/i_paid_4 _million_for_this_.html

26. http://www.fastcompany.com/magazine/164/major-league-baseball -advanced-media-bam

27. When Comcast acquired NBC Universal, it estimated the value of the broadcast television side of NBC at $0. Cable fees are threatened as more and more Americans "cut the cord," reducing their cable fees in favor of online alternatives. DVD sales have collapsed, down 25% since their peak in 2004. Box office revenues are at their lowest rate in history as Americans stop going out to the movies, thanks to better home theater technology.

28. http://rogerebert.suntimes.com/apps/pbcs.dll/article?AID=/20111228 /COMMENTARY/111229973

29. http://www.shortoftheweek.com/2012/01/05/has-hollywood-lost-its -way/

30. David Downs, "Five Kickstarter Projects Get Slammed with Success," *Wired* 20, no. 8 (2012): 25.

31. http://www1.umn.edu/news/news-releases/2012/UR_CONTENT _395531.html

32. Eli Pariser, *The Filter Bubble* (New York: Penguin Press, 2011).

33. Fred Hapgood, "The Media Lab at 10," *Wired*, March 2011.

34. http://mediadecoder.blogs.nytimes.com/2012/07/09/two-guys-made-a -web-site-and-this-is-what-they-got/

35. http://www.youtube.com/t/press_statistics/

36. It is worth noting that although streaming services like Spotify pro-vide an alternative to iTunes, some recent data suggests that these streaming services represent an even worse financial arrangement for musicians than iTunes. The royalties provided by streaming services are minuscule, about one tenth of one percent per stream. See http: //www.csmonitor.com/Innovation/Horizons/2011/0720/Spotify-Good -for-music-lovers-bad-for-musicians.

37. http://willvideoforfood.com/2011/07/04/youtube-earning-stats-partners -income/

38. http://www.thenation.com/article/168125/amazon-effect

39. http://www.wired.com/business/2010/06/apple-bans-cartoon-boobs -in-joyces-ulysses/

40. http://www.newyorker.com/reporting/2010/04/26/100426fa_fact_auletta

41. http://bits.blogs.nytimes.com/2010/01/29/amazon-pulls-macmillan -books-over-e-book-price-disagreement/

42. http://www.nytimes.com/2012/05/20/opinion/sunday/friedman-do -you-want-the-good-news-first.html

43. Qtd. in http://www.nytimes.com/2011/02/14/business/media/14carr.html

5. Big Government

1. Eamon Grennan, "Full Moon," *Still Life with Waterfall* (Minneapolis, MN: Graywolf Press, 2002), 54.

2. http://ofps.oreilly.com/titles/9780596804350/defining_government_2 _0_lessons_learned_.html

3. http://www.defense.gov/pubs/BSR_2007_Baseline.pdf; Chalmers Johnson, *The Sorrows of Empire: Militarism, Secrecy, and the End of the Republic* (New York: Metropolitan Books, 2004), 4.

4. http://www.wpri.com/dpp/news/local_news/providence/ri -house-to -vote-on-state-budget-plan; http://www.federaltimes.com/ article/20120723/DEPARTMENTS01/307230001/At-DoD-6-projects-124 -8-billion-over-budget?odyssey=mod%7Cnewswell%7Ctext%7CDepart ments%7Cp

5. Joshua Holland, *The Fifteen Biggest Lies about the Economy: And Everything Else the Right Doesn't Want You to Know about Taxes, Jobs, and Corporate America* (Hoboken: John Wiley, 2010), 105.

6. http://abcnews.go.com/blogs/politics/2012/01/congress-hits-a-new -low-in-approval-obama-opens-election-year-under-50/

7. http://abcnews.go.com/blogs/politics/2012/02/frustration-index-still -hot-in-the-kitchen/

8. http://pewresearch.org/pubs/1913/poll-trust-washington-anger-gov ernment-gay-marriage-support-abortion

9. Charlene Li and Josh Bernoff, *The Groundswell*, (Cambridge: Harvard Business Press, 2008).

10. http://www.wuala.com/en/bitcoin

11. http://www.launch.is/blog/l019-bitcoin-p2p-currency-the-most-dan
gerous-project-weve-ev.html

12. http://www.guardian.co.uk/media/2011/aug/08/london-riots-face
book-twitter-blackberry

13. http://articles.philly.com/2011-08-14/news/29886718_1_social-media
-flash-mob-facebook-and-other-services

14. http://www.csmonitor.com/USA/2011/0815/Flash-mobs-vs.-law-and
-order-BART-protest-adds-fresh-twist

15. http://www.nytimes.com/2010/03/07/magazine/07Human-t.html

16. http://gizmodo.com/5927379/the-secret-online-weapons-store-thatll
-sell-anyone-anything

17. "Businesses and others lobbying Congress reported spending $1.4 bil-
lion in 1998, a figure that more than doubled to $3.5 billion in 2009."
http://www.washingtonpost.com/wp-dyn/content/article/2011/02/02
/AR2011020205406.html

18. http://www.ajc.com/news/north-fulton/once-a-trendsetter-sandy
-903431.html

19. http://www.nytimes.com/2012/06/24/business/a-georgia-town-takes
-the-peoples-business-private.html?pagewanted=all

20. http://www.governing.com/blogs/bfc/sandy-springs-outsources-
everything.html

21. Bill Bishop, *The Big Sort: Why the Clustering of Like-Minded America Is Tearing Us Apart* (Boston: Mariner Books, 2009), 199.

22. "Hazards of Prophecy: The Failure of Imagination," in *Profiles of the Future:
An Enquiry into the Limits of the Possible* (1962, rev. 1973), 14, 21, 36.

23. On September 26, 2006, Facebook opened itself to anyone over the age of
thirteen with an e-mail address. Carolyn Abram, "Welcome to Facebook,
Everyone." The Facebook Blog (September 26, 2006).

24. Tim O'Reilly, "Government As Platform," in *Open Government: Collaboration, Transparency, and Participation in Practice,* edited by Daniel Lathrop,
Laurel Ruma. O'Reilly Media; February 23, 2010.

25. http://www.data.gov/sites/default/files/attachments/hbs_datagov_case
_study.pdf

26. http://www.census.gov/econ/smallbus.html

27. http://148apps.biz/app-store-metrics/

28. http://thehill.com/blogs/hillicon-valley/technology/101779-new-dc
    -cto-scraps-apps-for-democracy

29. http://www.cityofboston.gov/Images_Documents/02%20Summary
    %20Budget_tcm3-16341.pdf

30. http://www.thirty-thousand.org/

31. Tom Holland, *Persian Fire: The First World Empire and the Battle for the West* (New York: Anchor, 2007), 134.

32. http://articles.latimes.com/2009/may/28/opinion/oe-chemerinsky28

33. http://www.stjornlagarad.is/english/

34. http://w2.eff.org/Censorship/Internet_censorship_bills/barlow_0296
    .declaration

35. Gary Wolf, "Why Craigslist Is Such a Mess," *Wired*, August 24, 2009; http://www.wired.com/entertainment/theweb/magazine/17-09
    /ff_craigslist?currentPage=all

36. Ibid.

37. United Nations Convention Against Transnational Crime and Its Protocols. United Nations Office on Drugs and Crime, 2000. Retrieved from http://www.unodc.org/unodc/en/treaties/CTOC/index.html.

38. Paul Di Filippo, "Wikiworld," in *Fast Forward 1: Future Fiction from the Cutting Edge*, Lou Anders, editor, Pyr, 2007.

39. Alexis de Tocqueville, John Canfield Spencer, *The Republic of the United States of America: And Its Political Institutions, Reviewed and Examined*, 271.

40. Only 5 percent of registered voters in Dallas, 6 percent in Charlotte, and 7 percent in Austin turned out to vote in recent mayoral elections (http://www.fairvote.org/voter-turnout#.UA7sAUBYuZY).

41. http://student-of-life.newsvine.com/_news/2010/11/21/5502595
    -thomas-jefferson-supported-rewriting-the-constitution-every-19
    -years-equated-not-doing-so-to-being-enslaved-to-the-prior-genera-
    tion-what-do-you-think-about-that

42. http://www.huffingtonpost.com/howard-fineman/election-2012-lost
    -middle-consensus-internet_b_1502845.html

6. Big Armies

1. Adam Zagajewski, "Try to Praise the Mutilated World," trans. Clare Cavanagh, *Without End: New and Selected Poems* (New York: Farrar, Straus and Giroux, 2003), 60.

2. http://www.nytimes.com/interactive/2011/09/08/us/sept-11-reckon ing/cost-graphic.html

3. http://www.npr.org/templates/story/story.php?storyId=125186382

4. http://www.pbs.org/wgbh/pages/frontline/shows/front/special/tech .html

5. http://www.usatoday.com/news/nation/2010-08-25-1A_Awlaki25_CV _N.htm

6. http://www.nytimes.com/2009/11/10/us/10inquire.html

7. Qtd. in Andrew Hoskins, *Radicalisation and the Media: Legitimising Violence in the New Media* (New York: Routledge, 2001), 56.

8. http://www.foxnews.com/world/2011/05/15/bin-laden-logged-al-qaida/

9. http://www.washingtonpost.com/wp-dyn/content/article/2005/08/08 /AR2005080801018_pf.html

10. http://www.dailymail.co.uk/news/article-2155682/Al-Qaeda-posts -online-recruitment-adverts-offering-training-suicide-bombers-target -US-Israel-France.html

11. http://www.foreignaffairs.com/articles/61924/evan-f-kohlmann/the -real-online-terrorist-threat

12. http://www.theregister.co.uk/2008/09/12/lieberman_saves_youtube _and_world_from_terrorists/

13. http://www.youtube.com/t/community_guidelines

14. http://www.haaretz.com/news/international/report-bin-laden-hid-in -pakistan-compound-for-over-three-years-before-capture-1.359578

15. Qtd. in http://mediacdn.reuters.com/media/us/editorial/reuters-mag azine/reuters-aspen-2012.pdf

16. http://news.cnet.com/8301-31921_3-20062755-281.html

17. http://www.bloomberg.com/news/2011-09-09/terrorism-after-bin -laden-seeks-smaller-targets.html

18. http://gizmodo.com/5927379/the-secret-online-weapons-store-thatll -sell-anyone-anything

19. http://www.reuters.com/article/2012/05/23/us-arms-navies-smallships
-idUSBRE84M0PW20120523

20. http://www.armytimes.com/legacy/new/0-292925-1060102.php

21. http://krebsonsecurity.com/2011/01/ready-for-cyberwar/

22. http://www.brookings.edu/research/articles/2011/08/15-cybersecu
rity-singer-shachtman

23. http://www.washingtonpost.com/national/national-security/cyber
-intruder-sparks-response-debate/2011/12/06/gIQAxLuFgO_print.html

24. Seymour Hersh, "The Online Threat: Should We Be Worried about a
Cyber War?," *The New Yorker,* November 1, 2010; http://www.newyorker
.com/reporting/2010/11/01/101101fa_fact_hersh?currentPage=all.

25. Ibid.

26. http://www.wired.com/threatlevel/2010/04/cyberwar-richard-clarke/

27. www.youtube.com/watch?v=swHkpHMVt3A

28. Julian Assange, "The Non Linear Effects of Leaks on Unjust Systems of
Governance," *WikiLeaks* 31 Dec. 2006. Archived from the original on 2
Oct. 2007.

29. http://www.smh.com.au/technology/technology-news/the-secret-life
-of-wikileaks-founder-julian-assange-20100521-w1um.html

30. http://www.guardian.co.uk/world/2010/dec/01/us-embassy-cables
-executed-mike-huckabee

31. http://www.telegraph.co.uk/news/worldnews/wikileaks/8171269/Sarah
-Palin-hunt-WikiLeaks-founder-like-al-Qaeda-and-Taliban-leaders.html

32. http://www.guardian.co.uk/media/2010/dec/19/assange-high-tech
-terrorist-biden

33. http://www.democracynow.org/2010/12/31/pentagon_whistleblower
_daniel_ellsberg_julian_assange

34. Micah L. Sifry and Andrew Rasiej, *WikiLeaks and the Age of Transparency*
(Berkeley: Counterpoint, 2011), 14.

35. http://gov20.govfresh.com/samantha-power-transparency-has-gone
-global/

36. http://wikileaks.ch/cable/2008/06/08TUNIS679.html

37. http://www.time.com/time/magazine/article/0,9171,2044723,00.html;
http://www.cnn.com/2011/12/17/world/meast/arab-spring-one-year
-later/index.html

38. Rebecca MacKinnon, *Consent of the Networked* (Philadelphia: Basic Books, 2012), 57.

39. Evgeny Morozov, *The Net Delusion: The Dark Side of Internet Freedom* by (New York: PublicAffairs, 2011), 53.

40. http://www.fastcoexist.com/1678169/afghanistans-amazing-diy-internet

41. To this trinity, Morozov adds a fourth hallmark, provided to a large extent by the technologies of radical connectivity: entertainment. In a chapter titled, "Orwell's Favorite Lolcat" he writes, "Another Sakharov seems inconceivable in today's Russia, and in the unlikely event he does appear, he would probably enjoy far less influence on Russian national discourse than Artemy Lebedev, Russia's most popular blogger, who uses his blog to run weekly photo competitions to find a woman with the most beautiful breasts (the subject of breasts, one must note, is far more popular in the Russian blogosphere than that of democracy). . . . But efficiency and comfort—which the Internet provides—are not necessarily the best conditions for fomenting dissent among the educated classes. . . . If anything, the Internet makes it harder, not easier to get people to care, if only because the alternatives to political action are so much more pleasant and risk-free." The days of the master switch as a tactic might be numbered, but authoritarian regimes still have plenty of options: co-opt online communications with propaganda, distract through entertainment, or use social networking for surveillance.

42. http://arstechnica.com/business/2011/10/anonymous-takes-down-darknet-child-porn-site-on-tor-network/

43. http://www.vanityfair.com/business/features/2011/04/4chan-201104

44. Interview by Kim Masters and Renée Montagne, "Anonymous Wages Attack on Scientologists: The Fight Started When the Scientologists Tried to Get a Video of Tom Cruise off the Internet," *Morning Edition: Digital Culture* 7 Feb. 2008, National Public Radio.

45. http://boingboing.net/2011/05/29/pbs-hacked-in-retrib.html

46. http://www.guardian.co.uk/technology/2011/nov/02/anonymous-zetas-hacking-climbdown

47. Quinn Norton, a frequent *Wired* contributor on Anonymous, calls the group "a culture, one with its own distinctive iconography . . . its own

self-referential memes, its own coarse sense of humor. And as Anonymous campaigns have spread around the world, so too has its culture, bringing its peculiar brand of cyber-rebellion to tech-savvy activists in Eastern Europe, Asia, Latin America, and the Middle East. Like a plastic Fawkes mask, Anonymous is an identity that anyone can put on, whenever they want to join up with the invisible online horde" (http://www.wired.com/threatlevel/2012/06/ff_anonymous/all/)

7. Big Minds

1. William Butler Yeats, "Lapis Lazuli," *The Collected Poems of W. B. Yeats* (New York: Scribner, 1996), 294.
2. http://theamericanscholar.org/the-disadvantages-of-an-elite-education/
3. Derek Bok, *Our Underachieving Colleges: A Candid Look at How Much Students Learn and Why They Should Be Learning More* (Princeton: Princeton UP, 2008), 34.
4. Harry Lewis, *Excellence Without a Soul: How a Great University Forgot Education* (New York: PublicAffairs, 2006), 8.
5. http://www.usnews.com/opinion/articles/2011/03/18/a-harvard-education-isnt-as-advertised
6. http://www.demos.org/publication/great-cost-shift-how-higher-education-cuts-undermine-future-middle-class
7. http://www.washingtonpost.com/opinions/when-it-comes-to-e-mailed-political-rumors-conservatives-beat-liberals/2011/11/17/gIQAyycZWN_story.html
8. http://www.csmonitor.com/The-Culture/Family/2012/0617/Bachelor-s-degree-Has-it-lost-its-edge-and-its-value
9. http://chronicle.com/blogs/innovations/why-did-17-million-students-go-to-college/27634
10. http://pewresearch.org/pubs/1993/survey-is-college-degree-worth-cost-debt-college-presidents-higher-education-system
11. http://pewresearch.org/pubs/1993/survey-is-college-degree-worth-cost-debt-college-presidents-higher-education-system
12. http://www.businessweek.com/technology/content/may2011/tc20110524_317819.htm

13. http://techcrunch.com/2011/04/10/peter-thiel-were-in-a-bubble-and
    -its-not-the-internet-its-higher-education/
14. http://ocw.mit.edu/about/newsletter/archive/2011-10/
15. http://techcrunch.com/2011/10/19/khan-academy-triples-unique-users
    -to-3-5-million/
16. http://blogs.reuters.com/felix-salmon/2012/01/31/udacitys-model/
17. http://www.britannica.com/blogs/2008/04/the-great-unbundling
    -newspapers-the-net/
18. http://www.theatlantic.com/technology/archive/2012/01/the-great
    -unbundling-of-the-university/251831/
19. http://www.centerforcollegeaffordability.org/uploads/ForProfit_Higher
    Ed.pdf
20. http://www.bloomberg.com/news/2011-10-19/apollo-fourth-quarter
    -profit-sales-top-analysts-estimates-1-.html
21. http://nber.org/papers/w18201
22. http://www.theatlantic.com/business/archive/2012/07/why-the-internet
    -isnt-going-to-end-college-as-we-know-it/259378/
23. http://toolserver.org/~daniel/WikiSense/Contributors.php?wikilang=en
    &wikifam=.wikipedia.org&grouped=on&page=Abraham_Lincoln
24. http://storify.com/jcstearns/50-years-after-the-vast-wast
25. http://www.nytimes.com/2012/01/17/science/open-science-challenges
    -journal-tradition-with-web-collaboration.html?pagewanted=all
26. http://www.theatlantic.com/health/archive/2011/10/publication-bias
    -may-permanently-damage-medical-research/246616/
27. http://usefulchem.wikispaces.com/
28. David Weinberger, *Too Big to Know: Rethinking Knowledge Now That the Facts Aren't the Facts, Experts Are Everywhere, and the Smartest Person in the Room Is the Room* (New York: Basic Books, 2011), 139.
29. Ibid., 140.
30. http://blogs.discovermagazine.com/badastronomy/2007/07/11/discover
    -new-galaxies/
31. http://www.hbs.edu/research/pdf/07-050.pdf
32. http://www.parade.com/hot-topics/2008/09/secrets-of-great-presidents
    _233. http://lisaneal.files.wordpress.com/2009/02/alt12-gualtieri1.pdf
34. http://www.thedailybeast.com/newsweek/2012/04/15/why-your-doctor
    -has-no-time-to-see-you.html

35. http://papers.ssrn.com/sol3/papers.cfm?abstract_id=1549444

36. Weinberger is well-known for coauthoring the *Cluetrain Manifesto* (1999), a list of ninety-five theses that hammered away against the orthodoxy of the corporate establishment of the 1990s, especially in marketing and advertising. The first thesis states, "Markets are conversations" and is an explicit affront to traditional marketing and advertising that operate in a one-way "broadcast" mode.

37. http://www.theatlantic.com/technology/archive/2012/01/what-the-internet-means-for-how-we-think-about-the-world/250934/

38. Weinberger, *Too Big to Know*, 117.

39. "Smears 2.0," *Los Angeles Times*, 3 Dec. 2007, http://a.nicco.org/tcXn9b.

40. "Clinton Campaign Volunteer Out Over False Obama Rumors," *Washington Post*, 5 Dec. 2007, http://a.nicco.org/sBHCBm.

41. "N.Y. Mayor Urges Jewish Voters to Denounce Obama Muslim Rumors," *Associated Press*, 2 June 2008, http://a.nicco.org/uxPYNB.

42. http://www.washingtonpost.com/opinions/when-it-comes-to-e-mailed-political-rumors-conservatives-beat-liberals/2011/11/17/gIQA yycZWN_story.html

43. Michael Grant, *History of Rome* (New York: Charles Scribner, 1978), 264.

44. Office of Highway Policy Information, Table HM-20: Public Road Length, 2010 report (Federal Highway Administration, Dec. 2011).

45. http://www.gallup.com/poll/114544/Darwin-Birthday-Believe-Evolution .aspx

46. Colin Wells, *Sailing from Byzantium: How a Lost Empire Shaped the World* (New York: Bantam Dell, 2006), xxviii.

47. http://www.poetryfoundation.org/poetrymagazine/article/242902

48. T. S. Eliot, "Traditional and Individual Talent," in *Modernism: An Anthology of Sources and Documents*, Vassiliki Kolocotroni, Jane Goldman, Olga Taxidou, eds. (Chicago: University of Chicago Press, 1999), 369.

8. Big Companies

1. Philip Levin, "What Work Is," *What Work Is: Poems* (New York: Alfred A. Knopf, 1991), 18.

2. http://www.kauffman.org/newsroom/u-s-job-growth-driven-entirely
   -by-startups.aspx

3. Bureau of Labor Statistics

4. http://www.democracyjournal.org/20/individual-age-economics.php
   ?page=1

5. http://articles.economictimes.indiatimes.com/2011-05-18/news
   /29556184_1_financial-crisis-memory-east-india

6. http://www.democracyjournal.org/20/individual-age-economics.php
   ?page=2

7. http://www.fastcompany.com/33851/free-agent-nation

8. http://blogs.hbr.org/fox/2011/03/its-a-free-agent-nation-except.html

9. http://venturebeat.com/2008/01/10/27-electric-cars-companies-ready
   -to-take-over-the-road/

10. http://blogs.hbr.org/cs/2012/03/the_commoditization_of_scale.html

11. For Facebook, see http://www.datacenterknowledge.com/archives/2012
    /02/02/facebooks-1-billion-data-center-network/; for Google, see http://
    www.datacenterknowledge.com/archives/2011/04/15/google-invests
    -890-million-in-data-centers/.

12. http://www.nytimes.com/2012/05/05/business/deweys-collapse
    -underscores-a-new-reality-for-law-firms-common-sense.html
    ?pagewanted=all

13. http://www.newgeography.com/content/002977-the-rise-the-1099
    -economy-more-americans-are-becoming-their-own-bosses

14. http://www.freelancersunion.org/new-mutualism.html

15. http://www.worldchanging.com/archives/002365.html

16. http://bits.blogs.nytimes.com/2011/11/13/disruptions-the-3-d-printing
    -free-for-all/

17. http://www.dailymail.co.uk/sciencetech/article-1362049/Makerbot
    -thing-o-matic-Photocopier-thatll-1-day-make-Stradivarius.html

18. http://www.forbes.com/sites/stevedenning/2012/01/31/is-the-us-in-a
    -phase-change-to-the-creative-economy/

19. http://www.worldchanging.com/archives/001141.html

20. Qtd. in http://www.worldchanging.com/archives/002142.html.

21. http://www.wired.com/wired/archive/13.09/fablab_pr.html

22. http://www.openfabrication.org/?page_id=401

23. Qtd. in http://www.openfabrication.org/?page_id=401.

24. http://www.wired.com/wired/archive/12.11/brands.html

25. http://blogs.forrester.com/jp_gownder/11-10-04-brand_loyalty_is_de clining_total_product_experience_chains_can_help

26. http://www.popularmechanics.com/cars/reviews/preview/local-motors -rally-fighter-off-road-test-drive

27. http://www.wired.com/techbiz/media/news/2005/05/67612 ?currentPage=all

28. http://en.wikipedia.org/wiki/Usage_share_of_web_browsers#Historical _usage_share

29. http://www.orionmagazine.org/index.php/articles/article/6491

30. http://www.businessweek.com/articles/2012-05-10/more-farms-vie -for-the-1-billion-spent-at-farmers-markets

31. http://www.indyweek.com/BigBite/archives/2011/07/18/crop-mob -raids-66-cities-nationwide

32. http://switchboard.nrdc.org/blogs/pbull/job_creating_solar_energy_is _o.html

33. http://www.csmonitor.com/USA/Politics/2011/1107/Bank-Transfer-Day -How-much-impact-did-it-have

34. http://www.nytimes.com/2009/12/17/fashion/17etsy.html?page wanted=all

35. http://www.npr.org/blogs/money/2011/01/28/131450313/the-friday -podcast-pietra-rivoli-s-t-shirt-travels

36. http://www.orionmagazine.org/index.php/articles/article/6491

37. Christopher Steiner, $20 per Gallon: How the Inevitable Rise in the Price of Gasoline Will Change Our Lives for the Better (New York: Hachette Book Group, 2009), 141.

38. http://www.npr.org/templates/story/story.php?storyId=129151987

39. Rachel Botsman, Roo Rogers, What's Mine Is Yours: The Rise of Collaborative Consumption (New York: HarperBusiness, 2010), xx

40. http://www.fastcompany.com/1747551/sharing-economy

41. Rachel Botsman and Roo Rogers, introduction, What's Mine Is Yours: The Rise of Collaborative Consumption (New York: HarperCollins, 2010), xvii.

42. http://www.rollingstone.com/politics/news/global-warmings-terrifying -new-math-20120719

9. Big Opportunities

1. Eamon Grennan, "Could This Be It?", *Poetry Magazine*, August 1998.

2. Barbara W. Tuchman, *The Guns of August* (New York: Presidio Press, 1964), 1.

3. Catrine Clay, *King, Kaiser, Tsar: Three Royal Cousins Who Led the World to War* (Walker & Company, 2007), 95

4. With apologies to Clay Shirky (http://www.shirky.com/weblog/2009 /03/newspapers-and-thinking-the-unthinkable/).

5. http://www.usatoday.com/news/health/medical/health/medical/treat ments/story/2011/08/Study-Americans-use-of-antidepressants-on -the-rise/49828766/1

6. http://www.cnbc.com/id/31545004

7. http://www.caranddriver.com/features/texting-while-driving-how -dangerous-is-it

8. In her book *Sleeping with Your Smartphone* (Boston: Harvard Business Review Press, 2012), the Harvard Business School professor Leslie Perlow details her stunning research into the lives of business professionals, 70% of whom admitted to checking their smartphone (and e-mail inbox) within an hour of waking up.

9. http://www.sciencedaily.com/releases/2010/05/100528081434.htm

10. Sherry Turkle, *Alone Together: Why We Expect More from Technology and Less from Each Other* (New York: Basic Books, 2011), 284

11. Ori Brafman and Rod Beckstrom, *The Starfish and the Spider: The Unstoppable Power of Leaderless Organizations* (New York: Penguin, 2006), 8.

12. Apologies to Andrew Rasiej.

13. Sorry, Dad.

14. Rebecca MacKinnon, *Consent of the Networked: The Worldwide Struggle for Internet Freedom* (Philadelphia: Basic Books, 2012), 129, 164.

15. Doc Searls, *The Intention Economy: When Customers Take Charge* (Cambridge: Harvard Business Review Press, 2012), 8

16. http://www.vanityfair.com/business/features/2011/04/4chan-201104

Afterword to the 2014 Edition

1. http://www.theguardian.com/world/2013/oct/21/snowden-leaks
-france-us-envoy-nsa-surveillance
2. http://a.nicco.org/18ZGFEB
3. Original source of the quote is Lord Acton in a letter to Bishop Mandell
Creighton in 1887: "Power tends to corrupt, and absolute power corrupts absolutely." But it was popularized by George Orwell in *Animal
Farm*.
4. http://m.npr.org/news/U.S./235421758?start=0

ACKNOWLEDGMENTS

This book would not have happened without my agent, Lorin Rees. He called me one day out of the blue and said, "You should write a book!" Years later, after much encouragement and a lot of gentle prodding, here we are.

My editor at St. Martin's Press, George Witte, has taken an enormous chance on me as a first-time writer. The whole team at St. Martin's Press is stellar, especially the indefatigable Laura Clark.

Seth Schulman has been my guide and companion in learning to write—here's to you, Seth! Nicandro Iannacci was my loyal research assistant, diving into all manner of arcana without complaint; I hope this book helps you find a job on graduation.

My brother, Peter Mele, served as my first reader, followed by my parents, Mary Hélène and Nick Mele, and other family members. What would

I have done without their support, from my brother's clever asides, my father's careful grammar, and my cousin Jesse Davidson's nerd cred? My aunt, Linda Mele Dougherty, has cheered me on with sauce, love, and a family passion for reading. My grandfather Peter Davidson and my late grandmother Josephine Davidson were always book people, back to The Right Book Company. My in-laws have also been essential, starting with cousin Mary Aarons, who always has compelling, useful advice on book publishing. My mother-in-law, Pamela Aarons, has been a great support throughout, but in that final weekend of writing, she more than paid her dues by filling in with some last-minute kid watching.

Michael Ansara, Gina Glantz, Tyler Bridges, Zachary Tumin, Joe Costello, Matt Stempeck, Ethan Winn, and Max Novendstern all offered useful, detailed comments as early readers. I am especially indebted to Jan Frel, Garrett Graff, Mark Perry, and Micah Sifry for their careful reading and thorough feedback. A well-timed word from Ed Walker proved pivotal. I am deeply grateful to Dave Winer for his long mentorship and friendship; he taught me to think deeply about technology and to challenge my assumptions. Judge Mark Wolf unknowingly sparked the beginnings of this book by inviting me to join an argument on a cold day one November.

My colleagues at the Joan Shorenstein Center on the Press, Politics and Public Policy at the Harvard Kennedy School have challenged and encouraged me throughout, especially my boss Alex Jones. John Wihbey has been a great friend and helped me in many ways with the research for this book. I must also thank my students over the last few years at HKS; this book originated in many classroom conversations and arguments. A special word of thanks goes out to Nancy Palmer. Without her support, including detailed readings and corrections, this would have been a much lesser book.

My colleagues and clients at EchoDitto have played an integral role in the process. Lindsay Garber joined the team late in the process but is essential. Thank you! Liz Schwartz, who has filled almost every role imaginable, from scheduler to research assistant to proofreader to marketing manager, gets a special shout-out. Jon Haber has been a friend and mentor, as has Andrew Raisej. Joe Trippi—I owe you for the original subtitle, and for so much more. My two long-suffering business partners, Joshua Wachs and Justin Pinder, have taken many risks to make this book happen, and for that I am grateful.

And then there were all the people who took time out of their day to let me interview them, including but not limited to: Adrian Arroyo, Kieran Brenner, Ian Bogost, Rodney Cocks, Anthony DeRosa, Andy Eggers, Pete Forsyth, Katie Goodwin & Chris Francis, Neal Gorenflo, Brian Halligan, Ben Kaufman, Matt Macdonald, Nathaniel Pearlman, Bre Pettis, Hillary Rosen, Doc Searles, Michael Silberman, Maria Thomas, Mark Walsh, and David Weinberger.

A preemptive thank you to Dick Auletta and his team at R. C. Auletta and Company for their public relations work. Carolyn Monaco and her exceptional team at Monaco Associates, especially Christine Fadden, were essential in getting us focused on marketing and helping me navigate the world of book publishing. And a special thanks to Lindsay Tucker for her hard work on, well, everything.

It is hard to express the enormity of my gratitude for all the friends, family, clients, colleagues, students, and random acquaintances who have encouraged me from beginning to end, put up with my questions, and suffered my shortcomings, especially as my deadline loomed.

I have saved the best for last. My beloved wife, Morra Aarons—what can I say? You endured many nights and weekends as a single parent to see this book through, not to mention the angst, misery, and moods I carried with me along the way. You are my hero, my great love, and with any luck, you will let me write another book. (Did I push it too far?)

INDEX

PERMISSIONS